THE COMMON AGRICULTURAL POLICY AND THE WORLD ECONOMY

Essays in Honour of John Ashton

Edited by

Christopher Ritson

and

David Harvey

*Department of Agricultural Economics
and Food Marketing
University of Newcastle upon Tyne
UK*

C·A·B International

C·A·B International
Wallingford
Oxon OX10 8DE
UK

Tel: Wallingford (0491) 32111
Telex: 847964 (COMAGG G)
Telecom Gold/Dialcom: 84: CAU001
Fax: (0491) 33508

British Library Cataloguing in Publication Data
The Common Agricultural Policy and the world economy:
    in honour of John Ashton.
    1. Agriculture. Policies
    I. Ritson, Christopher, *1945 –*   II. Harvey, David III.
    Ashton, J. (John, *1922–1986*)
    338.184

ISBN 0–85198–688–9

Printed and bound in the UK

CONTENTS

iii

CONTRIBUTORS

Christopher Ritson
Professor of Agricultural Marketing
Department of Agricultural Economics and Food Marketing
The University
Newcastle upon Tyne NE1 7RU

David Harvey
Professor of Agricultural Economics
Department of Agricultural Economics and Food Marketing
The University
Newcastle upon Tyne NE1 7RU

Andrew Fearne
Lecturer
Department of Agricultural Economics and Food Marketing
The University
Newcastle upon Tyne NE1 7RU

Charles W. Capstick
Deputy Secretary of the Food Safety Directorate
Ministry of Agriculture, Fisheries and Food
Whitehall
London SW1A 2HH

Kenneth J. Thomson
Professor of Agricultural Economics
Aberdeen School of Agriculture
University of Aberdeen
Aberdeen AB9 1UD

Martin Whitby
Reader in Rural Resource Development
Department of Agricultural Economics and Food Marketing
The University
Newcastle upon Tyne NE1 7RU

Caroline Saunders
Lecturer
Department of Agricultural Economics and Food Marketing
The University
Newcastle upon Tyne NE1 7RU

Alan Swinbank
Professor of Agricultural Economics
Department of Agricultural Economics and Management
University of Reading
Whiteknights Road
Reading RG6 2AR

Simon Harris
Director for Corporate Affairs
British Sugar PLC
PO Box 26
Oundle Road
Peterborough PE2 9QU

Allan Buckwell
Professor of Agricultural Economics
Department of Agricultural Economics
Wye College (University of London)
Nr. Ashford
Kent TN25 5AH

John Lingard
Senior Lecturer
Department of Agricultural Economics and Food Marketing
The University
Newcastle upon Tyne NE1 7RU

Lionel Hubbard
Lecturer
Department of Agricultural Economics and Food Marketing
The University
Newcastle upon Tyne NE1 7RU

Tim Josling
Professor
Food Research Institute
Stamford University
Stamford
California 94305. USA

PREFACE

This book arose as a consequence of a series of public lectures on the Common Agricultural Policy, held at the University of Newcastle upon Tyne in honour of John Ashton, who died in July 1986.

John Ashton had been Professor and Head of Agricultural Economics at Newcastle for 24 years. He was previously Head of Economics in the Ministry of Agriculture, Fisheries and Food, London.

The lectures, which were all given by individuals for whom John had once been Head of Department, either at Newcastle or in the Ministry, were well received and well attended. We decided that it would be a fitting tribute to John to invite a number of his other former friends and colleagues to join the lecture-givers in contributing chapters in a book, dedicated to his memory.

John Ashton was universally liked and respected. A great internationalist, he was a founder member and Past President of the European Association of Agricultural Economists, a President of the British Agricultural Economics Society, and a strong supporter of the International Association of Agricultural Economists - a familiar face at conferences throughout the world. For years he was, internationally, perhaps the best known British agricultural economist. Many would argue that his greatest academic achievement can be found in the work he inspired in others, and we hope that this book may, in some small way, reflect that achievement. Certainly, it reflects the affection for John shared by its authors.

The editors would like to thank all who have contributed and, in particular, Helen Campbell, who was for many years secretary to John Ashton, and who has been responsible for the production of this book.

Christopher Ritson
David Harvey

December 1990

CHAPTER 1

INTRODUCTION TO THE CAP

Christopher Ritson

This short introductory chapter attempts only a very brief description of some central features of the operation of the Common Agricultural Policy. Some of the authors have assumed that readers will be familiar with the 'basic mechanics' of the Policy, but these are not dealt with explicitly in any of the other chapters. The chapter concludes with an outline of the way the book has been structured to provide what we believe to be a comprehensive and informed account of contemporary aspects of the European Community's Common Agricultural Policy.

THE TREATY OF ROME

The Treaty of Rome, which established the European Economic Community (EC) in 1958, required that 'The Common Market shall extend to agriculture and trade in agricultural products' (Article 38) and stated that 'The Community shall be based upon a Customs Union...' (Article 9). A Customs Union [1] is a form of economic integration in which all barriers on trade between member states are removed, and a common barrier is established on trade with third countries.

It was clear from the outset that this would require a **common** policy for agriculture. All member states had adopted complicated mechanisms for controlling agricultural product markets. The simple procedure of removing barriers on intra-Community trade, and establishing a common customs barrier on imports of agricultural products from outside the EC, would have involved developments in market prices for farm products which would have been unacceptable to most of the six member states.

Article 39 of the Treaty of Rome specifies a set of objectives for the Common Agriculture Policy (CAP) which are similar to those adopted by most of the developed countries. The policy seeks:

---

1   The relationship between the Common Agricultural Policy and a Customs Union is explored in Chapters 9 and 15.

1

'(a) to increase agricultural productivity by promoting technical progress and by ensuring the rational development of agricultural production and the optimum utilisation of the factors of production, in particular labour;

(b) thus to ensure a fair standard of living for the agricultural community, in particular by increasing the individual earnings of persons engaged in agriculture;

(c) to stabilise markets;

(d) to assure the availability of supplies;

(e) to ensure that supplies reach consumers at reasonable prices.'

Arguably, this is the most cogent statement of objectives for agricultural policy ever made, and several of our authors have found it helpful to refer to them when reviewing a particular aspect of the CAP (for example, Chapters 7, 10 and 12).

What is perhaps unusual about the CAP, however, is the severity with which one objective - the desire to protect rural living standards - came to dominate the way the policy was implemented. In particular, support prices for the major farm commodities were set at levels well in excess of those at which supplies could normally be bought or sold on world markets. Partly as a consequence of this, European farmers have been supplying a growing proportion of domestic requirements. Table 1.1 gives some estimates by the European Commission of the extent to which CAP support prices exceeded those applying on world markets [2] and Table 1.2 indicates the way self-sufficiency in farm products has tended to increase over the past 30 years. The Policy now covers all the major agricultural products produced within the EC, with the exception of potatoes.

An alternative way of characterising the CAP is by its 'three principles' rather than its five objectives. These were formulated somewhat obscurely during the early 1960s, but it quickly became *de rigueur,* as far as some member states were concerned, that any reform of the CAP must not call into question the three principles upon which it was founded. They may be described as:

a) no barriers on intra-Community trade in farm products;

b) preference for EC supplies in intra-Community agricultural trade;

c) common financial responsibility for CAP policies.

It was originally envisaged that the CAP would have two arms of approximately equal weight - a market arm and a structural arm. In the event, the market policy has dominated, taking the major part of expenditure on the CAP. The policy is financed by a special section of the Community budget,

---

2 In 1981, the Commission discontinued publication of this table, either because of discontent over the method of estimating the 'world price', or embarrassment over the outcome (or perhaps both).

usually known by its French acronym FEOGA - the European Agricultural Guidance and Guarantee Fund. Guarantee expenditure is for the market policies; the Guidance section finances structural reform. Details of the way the CAP has developed and is financed are given in Chapter 3.

Table 1.1: Prices of Certain Agricultural Products in the EC as a Percentage of Prices on World Markets*

Product	Marketing Year						
	1968 /69	1970 /71	1972 /73	1974 /75	1976 /77	1978 /79	1979 /80
Common Wheat	195	189	183	107	204	193	163
Rice	138	210	115	81	166	157	131
Maize	178	141	143	106	163	201	190
White Sugar	355	203	127	41	176	276	131
Beef and Veal	169	140	112	162	192	199	204
Pigmeat	134	134	147	109	125	155	152
Butter	504	481	249	316	401	403	411
Skim-Milk Powder	365	na	145	139	571	458	379
Olive Oil	173	155	125	113	192	200	193

na = not available.
* Most of the figures in this table are calculated by deducting import taxes (paid on whatever quantity of produce is imported) from the minimum import price for the commodity in question. Thus the 'world price' is a rough estimate of the average price received throughout the year by exporters to the Community and this will not necessarily always be representative of the price at which the commodity is traded elsewhere in the world. For some products, the 'World Price' has been estimated by deducting subsidies paid on EC exports from the national support prices.
Source: Commission of the European Communities, *The Agricultural Situation in the Community* (Annual Reports).

POLICY MECHANISMS

The core of the Common Agricultural Policy is the set of mechanisms which attempt to control the markets for agricultural products within the European Community. These vary significantly from product to product, and have been subject to repeated modification, particularly during the 1980s. It is nevertheless helpful to an understanding of the operation of the CAP market

Table 1.2: Degrees of Self-Supply in Certain Agricultural Products (EC Production as a Percentage of EC Consumption)

Product	1956/60 Average	1972		1981/82		1985/86		1987/88
	EEC-6	EEC-6	EEC-9	EEC-9	EEC-10	EEC-10	EEC-12	EEC-12
Wheat	90	111	99	125	127	132	126	119
Maize	64	68	58	72	73	94	82	91
Rice	83	112	92	131	130	82	79	79
Sugar	104	122	100	144	na	135	129	127
Fresh Veg.	104	100	94	97	100	101	107	106
Fresh Fruit	90	87	76	81	85	83	87	85
Citrus	47	52	34	36	45	47	75	74
Cheese	100	102	102	108	107	107	na	na
Butter	101	124	106	123	122	133	na	na
Eggs	90	99	99	103	102	102	na	na
Beef and Veal	92	81	84	104	103	108	107	107
Pigmeat	100	99	100	101	101	102	102	103
Poultry Meat	93	100	102	110	110	107	104	106
Veg. Oils/Fats	19	31	na	na	na	40	47	63

na = not available

Source: Commission of the European Communities, *The Agricultural Situation in the Community* (Annual Reports).

policy to construct a simplified 'ideal' model of a CAP market support system, to which all the specific commodity market regimes approximate to a greater or lesser extent.

Figure 1.1 therefore describes the essential features of a typical CAP support system for a farm product. [3] Each year, the Council of Ministers sets a **target price** for the commodity in question This price is intended as a guide to producers and a reference point for the operation of the Policy. The main mechanism which ensures that internal market prices are kept near the target level is a **levy** on imports which varies in such a way that imported products cannot undercut the target price. This levy is calculated by reference to a minimum import price - sometimes known as the **threshold price** - which is set a little below the target price to reflect the cost of transport from port to internal market centre.

---

3   Chapter 12 describes the same model using conventional economic supply and demand analysis.

A tax (the levy), equal to the difference between the threshold price and the world market price, is then charged on imports. 'World price' usually means, in this context, the lowest price at which a consignment of produce is being offered at a particular port during some specified time period.

Figure 1.1:   Model of Typical CAP Support System

The import levy will, on its own, keep internal market prices near the target level as long as the Community is less than self-sufficient in the commodity. If, however, EC farmers supply more of the commodity than can be sold on domestic markets at the target price, internal prices will begin to drop below target levels. For this eventuality, a second line of defence is required to prevent excess supplies (known as 'surpluses') from depressing producer prices. An **intervention price** is set, somewhat below the target price. If the internal market price should now fall to the intervention level, official intervention agencies will buy produce offered to them at the intervention price. The agency will then store the produce and subsequently export it at a loss. Alternatively, private traders receive an export subsidy (known as a **refund** or **restitution**) equal to the difference between the intervention price and the world price, and in fact the bulk of surplus production is now disposed of in this way. For some commodities, other methods are used to dispose of surpluses. For example, both wheat and skimmed milk powder have been subsidised for use as animal feeding stuffs.

It is evident from Table 1.1 that for most agricultural commodities, for most of the time, CAP support prices have exceeded 'world' prices. However, during the 1973/74 commodity boom, some prices for agricultural products on world markets did move above CAP support levels, and this happened again with sugar during the winter of 1980/81. For commodities in surplus, the EC is able to restrain domestic price levels by imposing export taxes but, as a general rule, import subsidies have not been introduced into the CAP system.

GREEN MONEY

In principle, the market support system should operate uniformly throughout the EC, with much the same import levy or export subsidy whichever member state trades with a third country, and with internal market prices, supported by the intervention arrangements, differing only on account of transport costs and adjustment lags. Support prices are fixed in a notional currency, originally called the Unit of Account (UA) and now the European Currency Unit. However, for the purposes of agricultural policy the European Currency Unit is deemed to have a higher value in terms of domestic and foreign currencies than it has for other purposes. The 'correcting coefficient', currently standing at 1.145109, means in effect that CAP prices are fixed in terms of a green ECU which has a value 14.5 per cent higher than the conventional ECU.

In order to maintain the unity of the internal market, prices should be converted into national currencies at prevailing market rates of exchange. This was indeed the case in the early years of the CAP, when rates of exchange between member states were fixed. But in 1969 the French franc was devalued and the German mark revalued, and since then, and particularly following the accession of the United Kingdom to the European Communities, there have been periodic changes in exchange rates and periods when some member state currencies have been allowed to float more or less freely against the others.

Fluctuating exchange rates raised the problem for the EC of what to do about CAP support prices in the member states involved. The maintenance of the unity of the market would have required support prices in domestic currency to be altered in line with the exchange rate change against the unit of account. But EC member states have often been reluctant to accept the full implications for domestic agricultural market prices of exchange rate movements. The reason is, in part at least, that the domestic implications for the prices of agricultural products under the CAP of a change in the exchange rates are much more abrupt, and more severe, than those for most non-agricultural products. In addition to this, the governments in Western Europe have become more committed to controlling the level of agricultural, and sometimes food, prices than they have for most other prices.

Because of this, member states, following depreciation or appreciation of their currencies, have been allowed to continue to convert CAP support prices into national currencies at the old conversion rate, which became known as the **green rate**. The effect is that, for a devaluing country, CAP support prices, and thus domestic market prices, are lower than implied by market rates of exchange, and for a revaluing country, higher. In order to maintain the price gaps it has been necessary to apply border taxes and subsidies on intra-Community trade - called monetary compensatory amounts (**MCAs**). In the absence of MCAs produce would flow from low-price to high-price countries (where it would have to be purchased by the intervention authorities) until the price gaps were eliminated. MCAs are also used to adjust import levies and export refunds on third country trade, thus reducing the net import tax (or export subsidy) in the case of the low-price countries, and raising them for the high-price countries.

The diagram illustrating the typical CAP support mechanisms therefore needs to be amended to take account of the price differences between member states. This is done, in a very simplified way in Figure 1.2, showing the situation for two countries, one which has experienced an appreciating currency ('Country 2') and one a depreciating currency ('Country 1'). (The diagram ignores the differences between the EC institutional prices - consolidating them into one 'EC support price'.) Periodic devaluation or revaluation of green rates have reduced or eliminated MCAs; but subsequently they reappear when market rates again diverge - though fixed positive MCAs are no longer permitted, which is what has led to the higher value of the 'green ECU' mentioned earlier.

STRUCTURE OF THE BOOK

We have divided the various aspects of the CAP considered in this book into four sections. Part one - entitled 'Understanding the CAP' - begins with a (deliberately) long chapter which traces the history and development of the Common Agricultural Policy from its origins in the 1950s through to the enlargement to include twelve member states in the mid-1980s. This is followed by three further chapters intended to provide a basic understanding of the Policy. Part two considers the relationship between the Common Agricultural Policy and five specific interests, namely, the consumer, the countryside, intra-EC trade, research policy, and the food industries.

Figure 1.2: Monetary Compensatory Amounts

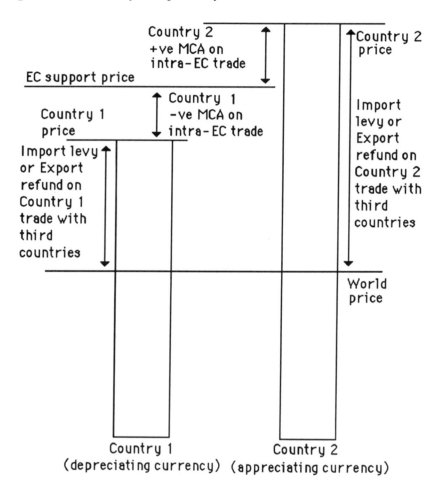

In part three we turn to the external face of the Policy. First, there is a general chapter on the impact of the CAP on world trade. This is followed by three chapters which look at the relationship between the CAP and, respectively, developing countries, the United States, and the Mediterranean countries. Throughout this section, frequent reference is made to the conflict between the trade impact of the CAP, and the attempt to liberalise world trade in agricultural products under the General Agreement on Tariffs and Trade (GATT).

The structure of the book was conceived before the dramatic events in Eastern Europe which gathered pace towards the end of 1989. We decided against the inclusion, at a late stage and simply for cosmetic reasons, a chapter on the CAP and Eastern Europe. But we recognise that this will represent an important new dimension to the EC's agricultural trading relations as the 1990s progress.

The final part, called 'The CAP and the Future', looks, first, at how the CAP has always been subject to pressure for reform, as a guide to how the policy is likely to develop in the future. Finally, we look at one specific possible development of the Policy - known as 'decoupling' support for the agricultural sector from support for farm product prices.

First, however, we reprint (with permission from the *European Review of Agricultural Economics*) John Ashton's 1984 Presidential Address to the European Association of Agricultural Economists.

CHAPTER 2

AGRICULTURAL MARKETS AND PRICES

Presidential Address to the IVth European Congress
of Agricultural Economists, Kiel, Germany. 3-7 September 1984

John Ashton

It is with great pleasure that I welcome delegates and other visitors to the Fourth Congress of our Association. I am privileged to have been President of the Association during the past three years and this Congress marks the conclusion of my period of office. I am sure that the Congress will be fruitful and thought-provoking and that everyone will emerge both intellectually and physically refreshed at the end of their stay in this delightful setting. The local organising committee has done an excellent job and I would like to express my thanks to them for it in advance of the official votes of thanks that will be made at the end of the week.

It has also been my privilege to have been closely involved with the development of the European Association. There was continuing dialogue among some of those involved in agricultural economics in Europe for a number of years before the first, tentative, congress at Uppsala, in Sweden in 1975. The outcome was that, at Uppsala, the Association was founded and its constitution and objectives defined. Following that, we had the Congress at Dijon in France in 1978 and then at Belgrade in Yugoslavia in 1981.

Apart from the triennial congress, the Association has developed specialist symposia which have taken place about twice a year. The organisation is undertaken by individuals and institutions, but these meetings have been treated as EAAE events with appropriate publicity given in the Newsletter. It is important that we continue to strive for the highest academic standards at these two types of professional meetings and pay great attention to the quality of the programmes proposed.

In addition to these activities, the Association also has a wider role to play in helping to stimulate collaboration between members from different countries and in encouraging interchange of staff and students. There is also the need to foster links between our countries and, especially, between Eastern and Western Europe. Nor must the Association see itself as being only concerned with narrowly European matters. Europe is the largest import market for agricultural produce and is the second largest agricultural exporter. It is vitally important that we look at our interrelationships with the rest of the world, especially with the developing countries. It is equally important that we have regard for our relationship with the rest of the European economy. Despite its size in the economy, agriculture cannot isolate itself from forces that shape the process and development of society. Its linkage with the macro economy should always be borne in mind.

The Association provides a focal point and stimulus for the development of the agricultural economics profession in Europe. Many of us have benefited from graduate training in the United States and have gained much help and co-operation from our American colleagues over the years. I should like to see Europeans achieve a similar dynamism in the agricultural economics profession, and particularly by the movement of academics in and out of government service. European agriculture covers a unique range of economic, social and political problems and a strong Association of Agricultural Economists should provide the locus for Europeans to derive intellectual stimulation from each other as well as from wider international sources.

Perhaps I could talk for a short while about the themes for our congresses. At Uppsala the theme was 'Short-term Prospects for the Development of European Agriculture'. At Dijon it was 'European Agriculture in a More Integrated Economy'. At Belgrade, the theme was 'Agriculture and Regional Development in Europe'. These intentionally diverse themes have, in many ways, reflected the more immediate concerns of agricultural economists about the role of agriculture in industrial economies and rural development. However, more traditional topics continue to have a profound effect on the economic well-being of farmers in the agricultural community and on public expenditure: topics such as production economics and farm management, agricultural trade and, particularly, the theme of this conference - agricultural markets and prices.

In all countries in Europe, there has been a substantial departure from the neo-classical model of free and competitive agricultural markets and price formation. In the centrally planned economies the agricultural sector is planned in detail as an integrated part of the total economy. In the Western European countries there are varying degrees of intervention into agricultural prices and markets, often very substantial. Prices and markets provide important signals to producers and consumers with regard to the efficient allocation of resources within agriculture and to purchasing by consumers, yet our diverse systems of intervention commonly lead to substantial misallocation of resources. In particular, such misallocation occurs through the encouragement of economically inefficient investment decisions. Also, much of the benefit from the intervention is capitalised into input values, particularly into land prices and rents, thus benefiting the landowner's wealth at the expense of the farmer's income for which the support was intended.

Intervention measures take place by a variety of means, notably through regulated prices, but also through market organisations including co-operatives, marketing boards and other agencies of a public nature, and through various more or less concealed charges imposed on the consumer. Most members of the public, unless they have a specialist interest in the subject, are not particularly aware of these departures from the market system, or the costs imposed on them by these departures. Moreover, when intervention becomes supranational, as in the EC, the conventional checks and balances through the political process on the extent of intervention become diffused and less effective.

Nor are agricultural intervention and its attendant distortions confined to Europe. The rest of the industrial world and the developing countries generate similar if not worse problems. In many of the developing countries it is a more obvious problem because of the relatively simple structure of the economy and the dominance of agriculture in it. Their problem, of course, is often the reverse of the developed countries. That is to say, prices are too low to stimulate the supply response needed for adequate supplies, leading to imports and the use of scarce foreign exchange for imported food. In a recent issue of *The Economist* (1984) the situation was stated succinctly:

'Since 1945 by fixing food prices in rich countries artificially high and prices in poor lands artificially low, governments have contrived to keep food output down where it is most needed and keep it up where farmers could be doing other things more profitably. The result is that 2 per cent of the world's farmers - the 24 million in the rich countries where less than one tenth of people work on farms - are providing nearly a quarter of the world's food, and nearly three-quarters of its food exports. The other 98 per cent of farmers - the 12 billion in the poor countries where up to two-thirds of people are on the land - grow three-quarters of the world's food but still do not have enough to eat.'

The implications of this state of affairs with regard to farmers in developing countries were spelt out later in the same article:

'While most rich democracies are overprotecting the 5 per cent of their people who are farmers, most poor dictatorships are milking the 70 per cent of their subjects on the land. State marketing boards force farmers to sell their crops at less than world market prices. Governments maintain overvalued exchange rates which keep the cost of imported food artificially low. Their big idea was often to industrialise helter-skelter by investing the countryside's surplus in the towns. So peasants flocked to the town where food was cheap while those who remained behind retreated into subsistence farming.'

Moreover, it would appear that the role of government in agriculture has not progressively decreased. On the contrary, it is expanding and will expand even further. On equity grounds, incomes are supported, generally through the price mechanism, resulting in increasing production and gains in productivity. This, of course, confronts policy-makers with the dilemma of the trade-off between efficiency and equity objectives. Equally, there is a keen political awareness concerning the security and stability of food supplies. Now we have new factors influencing government intervention and new sets of arguments. Strident voices are being heard throughout the world about protection of the environment. There is widespread and growing concern about the health and nutrition both of consumers and of farm animals. In some respects, of course, these are countervailing lobbies stimulated by the power of the agricultural lobby itself. Part of their success in being heard may have been generated by the increasing concern during a period of wide-scale economic recession with regard to the cost of agricultural support measures. In time, the nutrition lobby can be expected to diminish as education and public information on food and health improve and eating habits change in society. Moreover, there are limits to how far politicians are willing to go when it comes to policies designed to curb freedom of choice. At the same time, the environment lobby appears well established and likely to go on developing as predominantly urban populations become better informed, more affluent and more mobile, and show a high preference for the public goods identified by the environmentalists.

The problem with intervention is that any initial intervention almost inevitably gives rise to problems which have to be dealt with by successive rounds of further intervention. The outcome is often a greater distortion of markets and further divergence from economic efficiency. Examples of this in the EC include the basic system of price support resulting in overproduction which in turn has led, in the case of milk, to quotas and retirement subsidies. Rather than acknowledge the central importance of price as a determinant of supply, politicians appear to turn in any direction when it comes to trying to restrict the latter, rather than adjust the price. The problem is that price is also a determinant of revenue and hence of income as well. Thus, politicians seeking to maintain incomes cannot afford the price cuts necessary to return a distorted market to a competitive equilibrium.

For the first half of my career, I was an agricultural economist in government service and able to observe the pressures on agricultural policy-makers. These are many and varied, and although at times policy decision-makers might seem biased in favour of farmers, there are other powerful pressures pulling in other directions. Thus traders and food manufacturers make their voice heard, as do consumers, landowners and finance ministries. I sometimes wonder, in fact, if agricultural economists generally are as aware as they should be of the diverse interests that influence agricultural policy, and that all of these interests are worthy of study by agricultural economists.

A curious new contrast is now becoming even more striking between the Eastern and Western European groups of countries. In the West, we are seeing the introduction of quantitative controls in order to cope with the public expenditure consequences of income support policies and the regulation of trade, both imports and exports. In the East European countries we are seeing the recognition of the role of markets in providing an incentive through the price system for growth of output and productivity gains. This has led to some decentralisation of decision-making outside the central planning process and a reliance on signals from the market-place. Thus, in effect, both groups of countries are applying mixed strategies of market forces and physical controls such as quotas.

Part of the general problem with agricultural policies in the market economies is that they have made little allowance for productivity growth, and yet we know that in Western Europe at least productivity growth has been in the order of 1.5-2.0 per cent per annum and I think it might well be at least as great in Eastern Europe. Pricing policies have very seldom recognised the gains to producers arising from such growth. Over a period of several years, these are very substantial and can have an enormous bearing on the future shape and development of the industry. Many of the gains are derived from public resources: investment in agricultural research, development and extension, as well as public support for agriculture. It is not unreasonable to suggest that the consumer and taxpayer should benefit from some of these longer-term developments in the industry, although some would argue that consumers already gain in times of inflation through the erosion of real food prices. Politically, the problem is that the landowner gains at the expense of the farmer, since these benefits are soon capitalised in land values and increased rents. Thus, even with relatively high support prices, farming income net of real rents and capital charges does not always reflect the growth in output value or physical productivity.

The Economist article, from which I quoted above, suggested that a more realistic attitude might be emerging in agricultural policy formation on an international basis. Various tentative examples were given suggesting there might be moves in this direction. Certainly, increasing emphasis has been given to the role of prices in developing countries and realistic prices are increasingly advocated in order to stimulate growth, productivity and increased exports. At present, any prospects for progress towards these correspondingly liberal objectives in developed countries look, to me, rather bleak. But it may well be that the new realism is emerging with regard to developing countries. This development is in contrast to that of fifteen years or so ago when there was considerable momentum generated by UNCTAD and its supporters with the concept of a New International Economic Order. In fact, the momentum faltered and eventually it has almost disappeared from the agenda, with the Brandt Report, I suppose, as its epilogue.

Instead, we are now beginning to see the identification of specific problems which are analysed and solutions recommended. For example, there is currently much concern about the rescheduling of the debts of developing countries where payments and service charges constitute such a high proportion of export earnings and the foreign exchange available. Second, there is growing concern with the distorting effects of food aid, giving rise to dangers of long-term dependence on aid and a failure to allow domestic prices to stimulate increased domestic production in the countries concerned. Third, there is equally growing concern in traditional exporting countries with the disruption to markets and the depression of prices which are associated with the exports of surplus production from developed countries, notably the European Community.

15

In all of these areas, there is much scope for empirical research by agricultural economists. Market performance and the efficiency of market channels is a much neglected area within the general span of our research studies. In addition, there is only a limited study of the efficiency of marketing organisations, co-operative marketing boards and so on, many of which are of a state or parastatal nature, or else regarded benignly by governments. In this category, I would also include finance and credit organizations which have tended to be under the shelter of government protection, very often exempted from normal economic and commercial constraints affecting the rest of the economy.

Next, the area of commodity studies of an empirical nature is one in which there is vast scope for more detailed research. The trouble with relatively fixed price regimes, where prices are underpinned by guarantees, is that the price mechanism is not allowed to do its job as a regulator of the market. Nevertheless, variations in market price under such systems do allow scope for examining both the supply and demand response to price change and also the potential effect of production outside Europe on world markets.

Another area for general research and investigation is that of consumption. Generally speaking we have experienced, and will in the future experience, some degree of economic growth. This has a profound effect in the longer term on the habits of consumers. Not only will there be these induced changes in consumption, we now have added pressures arising from nutritional considerations of health and disease.

In mentioning these areas of potential research I have not been particularly precise and certainly not exhaustive. On the contrary, I have mentioned them by way of pointing to substantial gaps in our coverage of important research areas that could throw light on price formation and market behaviour. In so doing, such studies would provide a better foundation of analysis and fact on which policy-makers could make their decisions. Alternatively, such studies could help in pointing to the opportunity cost of departing from optimum solutions.

In talking about research, perhaps I should reiterate that agricultural economists are communicating with a wide variety of audiences, not just restricted to universities and agricultural ministries. There are the marketing organisations, private and co-operative, the food processors, manufacturers and agribusinesses, and of course the consumers and taxpayers. While in a congress such as this we are, in effect, talking to ourselves, it can also be said that we represent a wide range of interests with regard to our professional work and, of course, also fall into the last categories of audiences. I should add that, in looking through the detailed Congress Programme, I have been struck by the wealth and diversity of the contributions to our deliberations. It suggests to me that agricultural economics is in a healthy state and we have a good foundation for a promising and successful week of educational instruction and dialogue.

Judging from the attendance at this Congress, I would assume that the theme has generated wide interest. It is certainly the largest of our four congresses. Part of the reason is that information about the Association is now widely available. Part of it, I am sure, is due to the excellent location and arrangements made by our hosts. Nevertheless, I find it interesting that our Conference has attracted so many agricultural economists both from Europe and elsewhere to participate in the debate on our theme. We could well be on the threshold of a change in attitudes on levels of intervention and associated degrees of protection and on attitudes to food and agriculture. For young agricultural economists, this is an exciting time and perhaps one in which they may wish to realign their research priorities.

The development of links between agricultural economists in different countries in Europe existed, to some degree, before the Association was founded. However, the Association has helped to accelerate the interchange between agricultural economists in Europe and I hope it will continue to do so steadily and fruitfully, particularly with regard to some of the important research needs, including those I have just mentioned.

These developments can lead to healthy cross-fertilisation of ideas. Collaboration between universities and other institutions in the development of research projects can be mutually beneficial. A very limited amount of such collaborative research has taken place, but I hope that the Association will foster any opportunities for it that may be developed. Second, and I think there might be even more scope here, is the interchange of students. In the last decade or so there has been a certain degree of movement around Europe of undergraduates, postgraduates, and staff of agricultural economics departments. This is a matter greatly to be encouraged and facilitated. Certainly in my Department at Newcastle, we have benefited from visitors at all levels for periods varying from a few weeks to the full period of a degree course. We have also arranged for our own students to go to other European countries for varying periods. This has been immensely stimulating for all the students concerned. It has also been interesting for staff members to welcome students from different backgrounds and university traditions. These exchanges have given us in Newcastle great satisfaction.

Finally, I would like to say a word about the *European Review of Agricultural Economics*. Agricultural economics, like the industry it studies, is geographically diffused into many small compartments. Therefore, we must try to work together and an important instrument in the sort of co-operation needed is the *European Review* as a prime channel of communication. I would urge that we give it the maximum support possible, and hope that one day the relationship between ourselves and the *Review,* which is most cordial and collaborative, will be strengthened into some formal link so that the *European Review* becomes the official publication of the European Association.

We have come a long way from Uppsala in 1975. We still have a long way to go. Nevertheless, if the progress we have made in this Association is any guide to our future development, then the road looks relatively straight, direct and promising.

REFERENCES

Brandt Report(1980) *North-South : a Programme for Survival.* Pan World Affairs, Pan Books, London and Sidney.
The Economist, April 14, 1984.

PART I

UNDERSTANDING THE CAP

CHAPTER 3

THE HISTORY AND DEVELOPMENT OF THE CAP 1945-1985

Andrew Fearne

INTRODUCTION

This chapter is concerned with the stages leading to European integration and a common policy for agriculture. It discusses the influences emanating from the political parties and interest groups within the six original members of the Community, and the development of the CAP during a period in which the European Community (EC) grew from six to twelve member states.

THE ORIGINS OF THE CAP

1945-1954: Post World War II - the Preamble to European Union

When examining the factors which shaped the CAP from the outset, it is important to remember that a common policy for Western Europe's agricultural sector was the result of (rather than the reason for) a common desire amongst the nucleus of West European countries to establish a political and economic union.

Economic union was to be a means of political unification and a guarantee for peace following the resolution of world war. However, the impetus for economic and political co-operation came, not from within the Continent, but from the United States, which considered West European integration as the only way for Europe to recover from the aftermath of the war, and to construct an effective barrier to Communism and the expansion of the Soviet Bloc.

Greater co-operation was achieved in Western Europe through the implementation of the Marshall Plan, through which, between 1948 and 1952, some 25 billion dollars of aid from the United States and Canada was channelled. Even more important than this financial assistance (as far as European integration is concerned) was the emergence of the Organisation for European Economic Co-operation (OEEC), which formed the European

Recovery Programme (ERP) and established the priority of genuine European co-operation through the reduction of trade controls.

In the same year as the formation of the OEEC (1948), the elimination of customs duties and the institution of a common tariff for imports took effect in Belgium, the Netherlands and Luxembourg, under the agreement which formed the 'Benelux' Customs Union. The agreement, signed on September 5, 1944, called for co-operation in the field of tariff policies, which would lead to an economic union providing for '...the free movement of persons, goods, services and capital between the three countries' (De Vries 1975). While the Benelux partners could not have envisaged in 1944 the degree of economic integration which was to emerge in Western Europe, Benelux was clearly a forerunner of wider European union.

The impetus created by Benelux was carried still further, later in 1948, with the signing of the Treaty of Brussels, designed to encourage economic, social, cultural and defence co-operation. Significantly, Britain took part in this, along with France and the Benelux countries, but West Germany and Italy were not invited to sign the Treaty because those involved believed it was too soon after the War for active reconciliation with these two recent enemies.

A year later, following the Congress of Europe, the Council of Europe was created, involving a committee of ministers of national governments and an assembly of members of national parliaments. The Congress had adopted a resolution requiring the surrender of some national sovereignty prior to the establishment of economic and political union in Europe. The creation of a European parliamentary assembly, in which resolutions would be carried by majority vote, was proposed by the French, with the support of the Belgians. Britain was opposed to this form of supranationalism and managed to water down the original proposals considerably, with the Assembly and Council being granted no legislative powers but merely the role of a forum - a debating society designed only to provide recommendations to national governments (Swann 1981).

Britain's opposition to any form of supranationalism was reflected in her contribution to the first discussions on the common organisation of agricultural markets. A special committee was set up by the Council, in 1950, to examine the prospects for the integration of European agriculture. France was particularly keen to open up Europe's agricultural markets and proposed the creation of a 'High Authority' for agriculture, with extensive supranational powers; production was to be controlled, prices were to be fixed and all barriers to agricultural trade were to be removed. However, while Britain accepted the concept of an authority, it insisted on it being inter-governmental, with the modified role of reconciling differences in national agricultural policies.

As Tracy (1989) points out, these 'Green Pool' proposals were largely doomed from the start. The fifteen Western European countries which took part in the negotiations between 1952 and 1954 failed to reach any agreement, and Britain's opposition to the general idea of a supranational organisation was instrumental in this. However, as a further step towards European integration, the proposals and subsequent discussions served to identify the differences which existed between the countries involved, at least as far as agriculture was concerned, with France in particular committed to the 'European Movement' and Britain anxious to maintain her links with the Commonwealth and her sovereignty over policy formulation.

A much higher level of co-operation was established through the establishment of the European Coal and Steel Community (ECSC), in 1951. This time Germany and Italy were involved, along with France and the Benelux countries, but Britain did not participate. Again, the plan was essentially a French one, designed by Jean Monnet (then the head of the French State Planing Board) and based on the desire to integrate a rapidly reviving German economy into the rest of Western Europe while simultaneously ensuring that war between France and Germany would become not merely inconceivable but physically impossible (Kitzinger 1967).

The result, through the removal of customs duties, quotas, subsidies and price agreements, was the creation of a common market in coal, steel and iron, with a High Authority endowed with substantial direct powers. The High Authority was, however, responsible to a European Assembly, consisting of members elected by national parliaments, and on most matters the Council of Ministers representing member governments would have to be consulted. This was effectively the essence of the 'community method' of gradualist integration - progress towards political unity - by the integration of one sector at a time, with explicit political objectives being sought via economic co-operation. The 'Six' were in agreement, but the UK, although invited to join the Community, declined. With Monnet as the President of the first High Authority, the 'Six' were firmly set on a course to which the British were fully opposed.

The ECSC was an undoubted success with significant increases in the output and productivity of coal, steel and iron achieved from 1951 onwards. However, while the visible success of the Community lies in the impact on production, productivity and prices in the three industries, its real significance lies in the fact that it marked reconciliation between recent enemies and was the first grouping of the 'Six'. It also signified a stage in the recent history of the 'European Movement' when personalities came to play a major role in the composition and nature of an agreement designed primarily to further the federalist cause. Monnet and Schuman of France, Konrad Adenauer (the first Chancellor of West Germany), Alcide de Gasperi (Prime Minister of Italy) and Paul-Henri Spaak (Foreign Minister and former Prime Minister of Belgium), along with their political parties, were all agreed that nationalism should be contained while legitimate national and minority interests were safeguarded (Broad and Jarrett 1972).

The creation and subsequent success of the ECSC served to prove that the most effective solutions to the economic and political problems in Western Europe could be sought collectively, through the development of the European Movement.

1955-1957: the Treaty of Rome and Objectives for Agriculture

It was the Benelux countries who, aware of the difficulty of establishing a political union, eventually outlined a series of economic proposals for the creation of a fully integrated European market, covering a wide range of commodities. The memorandum which they presented recommended the convening of an inter-governmental conference to draw up treaties covering a general common market. Such a conference, involving the Foreign Ministers of the 'Six', but excluding an uninterested Britain, was set up in Messina, Sicily, in June 1955, where it was agreed to carry out the necessary preparatory work. The result of this was a report drawn up in 1956, under the chairmanship of Spaak, which formed the basis on which the Treaty establishing the European Community was built.

The chapter devoted to agriculture made it clear that the establishment of a common market in Europe which did not include agriculture was inconceivable (Comité Intergouvernemental 1956). However, the final resolution of the 'Six' to proceed with the construction of a European market included no specific reference to agriculture. As Neville-Rolfe (1984) points out, this omission was probably made to facilitate British participation in the steering committee, set up by Spaak. Britain did take part for a while, but within the year the British representative on the committee was withdrawn, with the Government publicly rejecting the proposals for economic unification and warning the six foreign ministers of the ECSC not to divide Europe by setting up an economic organisation separate from the OEEC. Nevertheless, the United Kingdom's withdrawal left the way open for the committee to draw up its proposals, published in April 1956, for a common market which included agriculture.

The Spaak Report outlined the special circumstances of European agriculture - the social structure of the family farm, the need for stable supplies and the problems resulting from climatic conditions and the inelastic demand for food. Moreover, it was recognised that the removal of tariffs and quotas (national intervention) would not be sufficient to allow the free movement of commodities between member countries. Those problems which demanded market intervention at a national level would not simply disappear with the creation of a common market. Thus, a common solution to the problems of European agriculture was sought (Fennell 1987).

To that end, the Spaak Report laid down a number of objectives for future agricultural policy, four of which were to be reflected in the Treaty of

Rome a year later: a) the stabilisation of markets; b) security of supply; c) the maintenance of an adequate income level for normally productive enterprises; and d) a gradual adjustment of the structure of the industry.

However, specific mechanisms for tackling these objectives were not detailed in the report. The long-term goal of a united Europe superseded those of a more temporal nature and few among those in favour of European integration wished to delay the creation of a common market while particular policy measures for agriculture were discussed.

National intervention in the agricultural sector varied among the 'Six' and, to avoid distortion, national policies should have been dismantled entirely. Had this been agreed, then competitive forces would have redistributed resources and changed the structure of agricultural production in Western Europe - an outcome totally unacceptable to the governments of the 'Six' (Marsh and Swanney 1980). Thus, when the heads of delegations met under the chairmanship of Spaak to negotiate the European Community Treaty, at Val Duchesse in 1957, the manner in which the decisions relating to agriculture were reached established that '...the common policy would be more a matter of accommodating national interests than of requiring radical adjustments' (Pearce 1983).

It is evident that during negotiations on the Treaty, agriculture was not considered a major priority. When the delegations of the 'Six' got together in Brussels to negotiate the Treaty, the working parties formed did not include one for agriculture. Indeed, in the Treaty itself, agriculture is only one of ten sectors in which the range of measures towards co-operation and integration were designed to apply. This lack of (agricultural) interest among the heads of government was largely due to the overwhelming desire for the creation of the Community not to be held up by wrangles over specific sectoral issues. For this reason, while the objectives for agricultural policy were detailed in the Treaty, the means by which they were to be achieved were not.

Not surprisingly, the Treaty itself deals more explicitly with the general goals of the Community, aimed at '...establishing a common market and gradually approximating the economic policies of the member states, to promote throughout the Community a harmonious development of economic activities, a continuous and balanced expansion, an increased stability, an accelerated raising of the standard of living and closer relations between the member states' (European Communities 1987).

Article 3 details how these objectives should be achieved: via the elimination of customs duties, quotas and so on; the establishment of a Common Customs Tariff (CCT) and a Common Commercial Policy (CCP) towards third countries; the removal of obstacles to the free movement of persons, services and capital, and the co-ordination of economic policies. Article 3 also refers directly to the creation of common policies for transport and agriculture. However, the member states were not obliged to complete a common policy for agriculture until the end of the twelve-year transition period, laid down for the achievement of the Common Market itself. Thus, it is not surprising that the explicit objectives described for agriculture (Articles 38-47) are fairly broad and allow for a range of interpretations.

Article 38 defines the scope of the Common Market as it applies to agriculture (covering products of the soil, of stock farming, of fisheries and of first-stage processing directly related to these products) and states that the Common Market for agriculture should be accompanied by a common agricultural policy. The objectives of the policy are set out in Article 39.1, as follows: a) to increase labour productivity by promoting technical progress and by ensuring the rational development of agricultural production and the optimum utilisation of the factors of production, in particular, labour; b) thus, to ensure a fair standard of living for the agricultural community, in particular by increasing the individual earnings of persons engaged in agriculture; c) to stabilise markets; d) to assure the availability of supplies; e) to ensure that supplies reach consumers at reasonable prices.

These objectives were to provide the yardstick by which all measures relating to agricultural policy would be judged acceptable, and indeed legal.

The vague reference to policy instruments is contained in article 40. It is here that the establishment of a market organisation is stipulated. The form of the organisation, however, was not detailed, but guidelines were provided. These included the regulation of prices, production and marketing aids, storage and carry-over facilities and the stabilisation of imports. Discrimination between producers or consumers within the Community was to be strictly avoided and any price policies adopted were to be based on common criterion and a uniform method of calculation. Article 40 also stipulated that a fund (or funds) should be created to finance the common organisation of agricultural markets.

The process by which the CAP was to be established was outlined in Article 43. The Commission was to submit proposals on the CAP to the Council of Ministers within three years of the signing of the Treaty and, following consultation with the European Parliament, the Council was required to make regulations, issue directives or take decisions thereon. Significantly, these regulations, directives or decisions were, for a 'short' period of time, to be made unanimously, with (qualified) majority voting to be established thereafter. This was an important 'federalist' input, aimed at removing the inevitable tendency for member states to think of nation before Community.

Recognition of the need to devote more time to the consideration of the machinery behind the CAP was also made explicit in Article 43, which recommended the convening of a conference between member states, to discuss their existing agricultural policies and formulate a statement of individual resources and requirements. If, as has been suggested, agriculture started out near the bottom of the signatories' list of priorities, this provision certainly confirmed that the consideration of agricultural problems and the construction of a common policy with which to tackle them, was to become a major factor influencing the structure of the Community during its infancy.

1958: The Stresa Conference - a Definitive Policy Framework

In accordance with article 43 of the Treaty, delegations from each member state, including (for the first time) representatives from the main farming organisations and the food industry, assembled at Stresa, Italy, in July 1958, to outline formally the problems to be tackled and the means by which they were to be resolved. Although the agreement reached at Stresa was not legally binding, the final resolution offered a more coherent view of the CAP than was presented in the Treaty of Rome.

Amongst the points agreed at the conference, the following were of particular importance. As mentioned in the Treaty, agriculture was to form an integral part of the overall economic strategy; trade was to be developed within the Community without threatening established political and economic ties with third countries; policies designed to manage the market were to be supported by structural measures, aimed at evening out production costs and ensuring a rational resource allocation, thereby stimulating productivity; equilibrium was to be sought between production and market outlets and it was hoped that increased productivity would allow the application of a price policy without the encouragement of over-production; aid to disadvantaged farmers was seen as a way of easing the necessary adaptations and a high priority was attached to increasing the efficiency of the family farm unit, which was to be safeguarded at all costs. Finally, it was hoped that the resultant improvement in the structure of the industry would enable capital and labour in the agricultural sector to receive remuneration comparable to that obtained in other sectors of the economy (Commission 1958b).

It was an impressive declaration of objectives and incentives for agricultural policy, but still lacking the precision which was a fundamental prerequisite to the eventual implementation of the CAP. This was largely due to the inevitable conflict of opinion which surfaced at the conference. However, the slick management of the conference by the secretariat of the Commission ensured that from conflict came compromise, and from ambiguity came cohesion. This in turn was largely the result of the determined effort from the Commissioner responsible for agriculture, Sicco Mansholt. He was to play a major role not only in the agreement at Stresa, but more significantly in the following years, during which time the CAP would officially come into being.

Like most of the representatives present at the conference, Mansholt was still in doubt over many of the key issues. He expressed his scepticism over the usefulness of price policies and his concern over the potential creation of surpluses and the effect this might have on the European Community's trading partners. But as far as the farming representatives were concerned, by far the most important issue which was covered was the principle that the family farm should remain the foundation of agriculture in the Community - a view to which Mansholt strongly subscribed and which he reiterated in the closing words of his final address: '...it is particularly encouraging that the conference has provided the opportunity for a frank discussion on doctrine and on the goals of our agricultural policy, that is to say, on the need to guide agriculture in the direction of sound family farms...In my view this must be so because...there can be no structural policy, or market policy, if we lose sight of this starting point, which in the long run is our final destination as well' (Commission 1958b).

Following the conference at Stresa, the Commission, in its first report on the activities of the Community, was to outline its views on the problems facing agricultural policy. Following the lines of the Stresa resolution, it considered the central problem to be the disparity existing between the level of income in agriculture and that in other sectors of the economy. The economic and political necessity of maintaining trade relations outside the European Community meant that the Community could not become a 'self-sufficing entity' and the Commission, like the delegates at Stresa, warned of the potential dangers of price support, stating that 'It would serve no useful purpose to ask for improvements in the structure of agriculture if prices were at the same time fixed at a level which enabled even those enterprises to cover their expenses, which owing to their inferior structure were producing at high costs' (Commission 1958a).

The Commission was clearly still not prepared to produce a blueprint for the CAP, particularly as it was still undecided over the specific policy instruments to adopt. Nevertheless, it was by this time evident that the Community was seeking to formulate a policy which would safeguard the family farm and support farm incomes, while simultaneously avoiding surpluses and maintaining trade links with third countries. The conflicting elements were still to be reconciled and the difficulties inherent in such a task meant that it was another four years before a policy sufficiently flexible to accommodate such diverse constraints could be established.

1959-1962: Conflict and Compromise - the Birth of the CAP

Following the Stresa Conference, the Commission gave itself a limit of two years from the Treaty's signature within which to submit proposals to the Council of Ministers on the working and implementation of the CAP. In November 1959 a draft of its general proposals was submitted to the Economic and Social Committee (ESC) and from March 1958 to December 1959 the European Parliament debated the issues raised. A final set of revised proposals were submitted to the Council in June 1960.

The proposals which the Commission presented to the Council, on the shape of the CAP, included the free circulation of agricultural products within the Community; progressive development over the transition period, in *harmony* with general economic and social activities; the close inter-dependence of structural, market and trade policies in agriculture; the eventual adoption of a system of common prices; and the encouragement and co-ordination of national policies in agricultural structure. The detailed proposals for market organisations for most of the main products rested on the principle of variable levies, both on third country imports and (during the transition period) intra-Community trade, with target and threshold prices as a means of harmonising existing policies

In July 1960 the Council met to consider the proposals and created the Special Committee on Agriculture (SCA), giving it a continuing mandate to prepare future Council decisions on agricultural issues. The proposals were further debated by the European Parliament in October and twelve months later, following negotiations between the SCA and the Commission, the Council eventually accepted the substance of the proposals for a system of levies, to be applied to intra-Community and third country trade. The Commission was then called upon to submit draft regulations applying the levy system to a series of products over the following year.

Throughout the year (1961) a number of draft regulations incorporating the mechanisms proposed by the Commission were circulated. These outlined the system of support prices, import levies and export refunds, intervention buying and so on, all of which were subsequently ratified. However, the Commission did not have it all its own way. For example, the idea of a central organisation responsible for calculating daily import levies, restitutions, and so on, was unacceptable, largely because member states wanted to keep intervention boards firmly under state control (Neville-Rolfe 1984). Equally unacceptable was the proposal for support policies to be financed by separate 'product stabilisation funds', with levy revenue providing the main source of finance, backed up (if necessary) by producer participation. The Commission realised, even at this early stage, that for certain products (notably milk) financial problems were likely and that making sectors accountable for their own financing would help to avoid this. However, most member states (particularly France) were in little doubt that such an idea would be impossible to 'sell' to their farmers and, as a result, no more was heard of producer participation in the financing of the CAP for a further fifteen years.

Following the draft regulations for cereals, pork, eggs, poultry, fruit, vegetables and wine, submitted in July 1961, the Council finally agreed, after more than two hundred hours of intensive negotiations, on January 14, 1962, to adopt a series of regulations giving legal effect to the levy system and instituting a common market organisation for each product. The levy system took effect from July 1, 1962 and from that date '...agriculture formally ceased to be a subject of purely national administration and control' (Lindberg 1963).

The elements of conflict, not only between member states but across policy objectives, meant that the Council had rejected the strict organisation of trade and markets based on quantitative restrictions, but they did agree on the three fundamental principles upon which European agricultural policy was to be organised: a) Market Unity - a single agricultural market, a common marketing system and common pricing; b) Community Preference - the competitiveness of Community producers should not be threatened by third country imports; and c) Financial Solidarity - expenses incurred to be financed by the Community, and income generated to form part of the Community's 'Own Resources'.

These three principles have (to varying degrees and with fluctuating emphasis) been adhered to throughout the CAP's existence and have been resolutely defended by the Commission (Ritson and Fearne 1984). Thus, while a number of issues were still to be resolved, notably the initial level of support prices, the agreement reached in 1962 undoubtedly signified the official birth of the CAP.

1963-1967: Common Prices and Common Financing - the Final Hurdle

To the extent that the objectives of the policy and the regulations outlining the market organisations for the various products were agreed in 1962, it can be said that that the CAP was 'created' in that year. However, before the policy could begin to operate effectively, let alone have an impact on the Community's agricultural sector, a number of crucial issues had still to be resolved: the level and seasonal scale of support prices, the size of import levies, the location of intervention centres, and quality standards.

Because of the inherent difficulties associated with establishing a common price level, progress on this issue was slow, with Germany in particular reluctant to adopt any form of common prices until the end of the transition period. However, pressure was brought to bear from outside the Community, in February 1963, when the Council asked the Commission to negotiate on the Community's behalf in the forthcoming GATT negotiations. Without a common internal price level it would have been difficult to adopt a common negotiating position on the level of prices. The most important product under consideration was grain, and with the United States and other grain-exporting nations eager to maintain trade flows with the European Community, internal price levels could not be set too high. It was important that a common price level should be agreed as soon as possible and the Commission began to apply pressure internally to this end, with a view to achieving the alignment of cereal prices to a common level by the beginning of the 1964/65 marketing year.

In the event, the deadline was not met and in October 1964 the French Government (under pressure from the French farm lobby) delivered an ultimatum to the Germans, threatening to withdraw from the Community if an agreement on cereals (crucial to France and the completion of the CAP) was not reached by the end of the year. Some concessions were forthcoming, notably the basic acceptance of a reduction in German cereal prices, to come into force in July 1967, but these were accompanied by demands for compensation and a revision of the price level in 1967, to account for the increase in production costs. This was unacceptable to both the French and the Italians and a deadlock seemed inevitable. However, on December 13 the German Minister for Economic Affairs, Schmucker, declared that his Government would accept the price reductions of between 11 per cent and 13 per cent, as proposed by the Commission, provided that they were not applied until July 1, 1967 (as opposed to 1966) and that they were accompanied by increased compensation and a revision clause. Three days later, after another marathon Council session, a final compromise was reached, largely due to French Agricultural Minister Pisani's decision to accept the proposed increase in compensation for German farmers. The Community had recovered from the brink of collapse and, following the agreement on cereals, the Commission set a similar deadline of July 1, 1967 for the alignment of national support prices for the other products. However, before celebrating the removal of the ultimate hurdle, the Commission still had to settle a number of issues involving the financing of the CAP.

There were two main problems with regard to the financial arrangements for the Community: the first concerned the allocation of national contributions to the Community budget during the interim period, before 'Own Resources' fully financed the Community budget; and the second referred to the deadline by which the Community was to be dependent upon those 'Own Resources'.

The Commission's proposals on common financing were submitted to the Council in March 1965. They concluded that the share of the budget attributed to the Guarantee and Guidance Fund (FEOGA) of the CAP should be increased, in order that total FEOGA expenditure could be met by the budget, from July 1, 1967. This meant that the single market should have been achieved in advance of the period laid down in the Treaty. In line with the regulations provided under the Treaty, the Commission proposed that the distribution of national contributions should be based on each country's share of agricultural imports from third countries. They also proposed that, under Article 201 of the Treaty, member governments should be asked to surrender their control over levy revenue and duties formerly accruing to national treasuries, and that this power should be granted to the European Parliament, thus increasing their control over the Community budget.

To this fourth proposal, the French Government was totally opposed, claiming that it was (at that time) unnecessary, and that it was in any case not the role designed for the European Parliament. At the same time Germany was against the proposed method of calculating national contributions on the basis that it discriminated against the net importing countries, of which it was one. Once again the positions of France and Germany were to provide the key to a solution. The French Government wanted an agreement over the method of financing the CAP, but refused to accept the strengthening of the Community's institutional powers. The conflict finally came to a head during the French presidency, when at the Council of Ministers meeting on June 28, the French Foreign Minister, Couve de Murville, left the Chair, at two o'clock in the morning of July 1, without an agreement over the finance issue. The French Government then announced it would take no further part in Council meetings.

It was six months before the 'Empty Chair' was reoccupied, when an extraordinary session of the Council took place at Luxembourg on January 17 and 18, 1966. With careful manipulation and interpretation of the constitution and regulations of the Treaty, a compromise was eventually found ten days later, when the meeting resumed. The financial arrangements were agreed and the allocation of national contributions was established on the basis of a 'fixed key', with ceilings for each country, applicable until 1970. This satisfied the Germans, who negotiated a ceiling of 31.4 per cent, compared with 32 per cent for France and 20.3 per cent for Italy. On the parliamentary issue, the debate was broadened to include the consideration of majority voting in the Council and the relations between the Council and the Commission. Thus, not only was it agreed to postpone ratification of the transfer of budgetary power until the end of the transition period, but the Council declared that in future, where 'very important interests' were at stake, a unanimous agreement would be necessary.

The 'Luxembourg Compromise' was a triumph for President de Gaulle, who foresaw the problems which could have arisen from the imposition of a majority decision contrary to the 'vital interests' of a member state. However, in establishing the means of protecting national interests, the pre-eminence of the Commission's proposals, envisaged by the Treaty of Rome, was substantially impaired. Nevertheless, as Tracy (1984) points out, the agreement did pave the way for work to resume on the remaining aspects of the CAP. Indeed, the agreement on common prices and the method of common financing was hailed as a considerable success, not only because the resolution (albeit temporary) of these problems allowed the effective implementation of the CAP, but also because it represented a considerable acceleration of the timetable laid down in the Treaty. As Raup (1970) succinctly puts it, '...the Common Agricultural Policy, which had been regarded as a bottleneck, became the 'motor' that drove forwards the total integration process and significantly expanded the area covered by community law'.

NATIONAL INFLUENCES ON POLICY FORMATION

During the post-war period, the nucleus of West European countries (governments and interest groups alike) were united in the belief that the political and economic problems which had emerged would be best resolved through co-operation and integration. Throughout the preliminary work, negotiations over the Treaty and subsequent agreements over the various policies adopted, one feature stood out clearly - the recognition of the need for a 'Community solution'.

The agreements reached inevitably reflected the extent to which national interests among the 'Six' conflicted. As far as the CAP is concerned, the nature and objectives of the policy established not only were a function of the contrasting agricultural policies and the relative importance of agriculture in the member states, but were equally dependent upon the distribution of political and economic power among the 'Six'. Thus, in order to gain further insight into the factors shaping the CAP and a better understanding of its importance within the Community, it is necessary to examine the agricultural situation and the type of policies operating in the member states prior to and during the policy-forming period.

France

The attitudes of the French Government and the French farm lobby were of crucial significance in the establishment of the CAP. Just as Germany played the leading role in directing the Community's industrial policies, so France was instrumental in ensuring the agreement on an agricultural policy which would enable her to offset (to a greater or lesser extent) the underlying weakness and uncompetitiveness of her industrial base.

Table 3.1 illustrates the dominant position of French agriculture in the Community during the policy-forming years: France accounted for over 45 per cent of the six countries' total agricultural area and around 40 per cent of total food production in the Community. It also had the lowest population density of the 'Six' and was by far the biggest exporter of agricultural products, particularly grain. Moreover, France was (and largely remains) particularly bound by agricultural tradition and, as a country at the heart of the European Movement, it is not surprising that the CAP was seen by many as a 'French Victory' (Clerc 1979).

There are a number of factors (geographic, demographic and political) contributing towards the important position of agriculture in the French economy. The strong concentration of industry in a few regional centres has meant that in large regions (notably Brittany and the Midi) agriculture provides the economic and thus political base. However, perhaps the most significant factor emanating from the structure of the industry is the fact that French farmers appear less committed than in other member states to a particular political party or doctrine (Tangermann 1980). Thus they constitute an effective body of floating voters who are consequently accorded political attention far in excess of their relative importance in the French economy, in terms of both population and economic activity.

Table 3.1: The Relative Importance of Agriculture in the 'Six'*

	France	Italy	Germany	Netherlands	Bel-Lux
Total arable area (sq.km)	346,330	209,650	143,320	23,100	18,720
Proportion of EC arable area (%)	46.7	28.3	19.4	3.1	2.5
No. employed in Agriculture (m)	3.7	5.0	3.0	0.4	0.2
Proportion of total EC workforce (%)	9.0	25.0	11.0	10.0	20.0
Proportion of EC food production (%)	39.4	26.3	23.0	6.4	4.9

* Rows 1 and 2 = 1965, rows 3 and 4 = 1964 and row 5 = 1955/56.
Source: Adapted from De la Mahotière (1970, p140) and Lindberg (1963, p272).

As far as the Commission's proposals on the formation of the CAP were concerned, French farm lobby groups in general reacted favourably. Eager to exploit their export capabilities, they insisted on a system based on Community Preference, with both levies and quotas on imports from third countries. Price reductions were not ruled out, providing safeguards were included to protect farm incomes while structural changes were made. Overall the attitude of French farmers was an opportunist one, best illustrated by a statement made by the vice-president of the Federation Nationale des Syndicats d'Exploitants Agricoles (FNSEA) - the main farm union organisation: 'Let us not forget the importance of the potential of French agricultural production, which must have markets. At our door there is a market of 170 million consumers which will become 180 million in ten years' time...It would be a grave error not to profit from the occasion and to let ourselves be put off by the obstacles that the Common Market will meet' (Lindberg 1963).

The securing of outlets for their agricultural surpluses was also a priority for the French Government. The four-year plan for 1962 to 1965 aimed at increasing agricultural output by 28 per cent and domestic consumption was not expected to keep pace. As a result, the French Government's stance reflected the position of the farm lobby and, as Lindberg

(1963) points out, subsequently became the closest of the 'Six' to that of the Commission. They insisted on the rapid implementation of the CAP and made this a condition for action on other areas (Neville-Rolfe 1984). The CAP was to be clearly 'preferential' and, while accepting the system of levies on intra-Community trade, the French argued for minimum prices, as a safeguard in 'exceptional cases'.

As it turned out, French concern over price levels was shown to be misplaced. Table 3.2 clearly illustrates the divergence of national farm support prices for the 'Six' and the position of France as the country with the lowest average support prices meant that French farmers would benefit from a common price level nearer the upper than the lower range of existing support prices.

Table 3.2: Average Wheat, Barley and Milk Prices in the 'Six' in 1958/59, 1966/67 and the Common Target Price Established in 1967/68 (UA per 100 kg)

	Belgium		France		Germany		Italy		Netherlands		EC (Target)
	58/9	66/7	58/9	66/7	58/9	66/7	58/9	66/7	58/9	66/7	67/68
Wheat	10.0	9.3	6.7	8.5	10.8	10.5	10.5	11.7	7.2	10.2	10.6
Barley	7.9	8.3	5.7	7.7	10.0	10.6	7.0	8.6	6.8	9.0	9.1
Milk	5.9	9.8	6.2	8.4	7.9	10.1	7.7	9.6	7.5	9.6	10.1

Source: Marsh and Ritson (1971).

So it was, and, as Kindleberger (1965) points out, the CAP was the main force behind the impetus which improvements in the French agricultural sector gave to the impressive overall economic performance of the French economy during the 1960s. Despite a considerable decline in the agricultural population and a reduction in the total cultivated area, output increased by over 50 per cent, the number of farms declined at an annual rate of 2.8 per cent per annum and the average farm increased in size from 20.4 hectares in 1963 to 27.6 hectares in 1970.

The effects of the CAP were there for all to see and, significantly, the gains made by the French from increased agricultural exports exceeded those made by Germany from industrial sales: between 1960 and 1966 French exports of foodstuffs to Germany increased in value from 142 million UA to 417 million UA, with no reciprocal advantage established in industrial goods (De la Mahotière 1970). However, while this confirmed French hopes of the impact which the CAP would have on intra-Community trade, it caused Germany to take a much closer look at the financing of the CAP, in particular the level of German contributions. The 1960s were clearly a period of accelerated growth in the French agricultural sector (and in the French economy overall), but the financial imbalances and growth of surpluses

35

which this expansion induced meant that the economic emphasis of the CAP would have to change over the following decade.

Italy

While the importance of agriculture was as great in Italy as in France during the post-war years (see Table 3.1), the influence of the Italian Government on the structure of the CAP was much less evident. This was largely due to the contrasting structure of Italian agriculture - self-sufficient in wine, olive oil and fruit and vegetables, but requiring substantial imports of meat, cereals and dairy produce. The CAP offered an extended market for Mediterranean products, from which the Italians could benefit, but high common prices for the 'Northern' products threatened to exacerbate the balance of payments problems which had long confronted the Italian Government.

Prior to the establishment of the Community, around 70 per cent of the country's total area was classed as agricultural land. Holdings were extremely small in size compared with the other five members and labour productivity was particularly low. However, as in France (though to a lesser degree), the proposition of the Common Market and the CAP 'shocked' Italian agriculture into accepting drastic structural changes, with much marginal land going out of agricultural use and the agricultural population falling by around 40 per cent during the 1960s, thus considerably improving the productivity of Italian agriculture, per man and per hectare (Parker 1979).

Despite the importance of agriculture in the Italian economy, there was a much weaker tradition of support for agriculture than elsewhere in the Community (Neville-Rolfe 1984). Apart from limited support for rice production and a high guarantee price for wheat, there was no system of direct aids to farmers and there were no duties or quotas on agricultural imports. Immediately after World War II, the Christian Democrat Government instituted a programme of extensive land reform. However, while this achieved little social or economic success, the political value of capturing widespread agricultural support for the Christian Democrats meant that further considerations of agricultural policy were effectively curtailed. The farm unions in Italy are more closely linked to the party political system than elsewhere in the Community and the land reform programme won the support of the Associazone Italiana dei Coltivatori Diretti (the largest farm union), representing some two million small-scale owner-occupiers who exercised a strong influence in parliament.

The Italian farming organisations in general were enthusiastic about the CAP and were the only allies of the French over the four main issues - common prices, the balance between structural and price policies, third country trade and the length of the accession period. Not surprisingly, they were especially concerned over exploiting their comparative advantage in the production of fruit and vegetables and pushed hard for the elimination of internal restrictions on trade in these products within the Community (particularly in France and Germany). However, as far as the other products were concerned, the Italian farm lobby was much more reticent about trade liberalisation and looked for extensive structural programmes as well as financial aid from the Community.

When the Italian Government did make its voice heard during negotiations on the CAP, its position also reflected its concern that Italian agriculture would find it difficult to respond to rigid price policies without structural measures to support the necessary adjustments. Thus, it was opposed to an accelerated harmonisation of prices on the grounds that Italian agriculture would need a longer period to adjust and become competitive. It also feared that price policies would merely lead to over-production of the 'Northern' products, the subsidisation of which would be part-financed by the Italian treasury, and thus demanded 'compensation' in the form of definitive structural programmes.

One feature which rendered the Italian Government's bargaining position even less persuasive during negotiations was their apparent contradictory attitude towards common prices. As Tangermann (1980) stresses, the high proportion of food imports and the economic significance of low food prices meant that the Italians were in favour of low support prices for some products (particularly the 'Northern' ones), while the desire to maintain agricultural employment and farm incomes, as well as to exploit its exporting capacity in fruit and vegetables, meant that the Italian Government would, on occasion, push for higher price levels.

The conflicting domestic interests over price levels meant that, rather than committing itself one way or the other, the Italian Government was forced to concentrate on increased financial aid towards structural programmes. However, during the policy-forming years, structural policy was not accorded high priority by the other member states.

The net result during the 1960s was that Italy did not benefit from the CAP as much as she had originally hoped. However, towards the end of the decade, in recognition of the less favourable (as far as Italy was concerned) aspects of agricultural policy embodied within the CAP, the Italian Government concentrated less on adopting a definitive position on price levels and placed more emphasis on negotiating specific measures in its favour. In this respect it was more successful, strongly influencing the decision to introduce support regimes for wine, tobacco and olive oil, and gaining preferential treatment over the distribution of structural funds.

The Netherlands

As the third major exporter of agricultural produce among the 'Six', the Netherlands, like France and Italy, was expected to benefit from the CAP. However, unlike the French and Italians, the Dutch did not depend so heavily on their agricultural sector as a foreign exchange earner, with Dutch industry during the 1950s and 1960s among the most efficient in Western Europe (De Vries 1975). For this reason, as well as her lack of political power, the Netherlands did not play such a significant role in the formation of agricultural policy within the Community.

The Netherlands is territorially one of the smallest countries in the Community, with its total land area barely 2 per cent of the Community's land area. However, as Table 3.1 shows, while occupying only 3 per cent of the Community's arable area in the mid-1960s, the Netherlands still managed to produce over 6 per cent of total European Community food production. With a total population representing only 5 per cent of the Community's but contributing around 6 per cent of its GDP, these figures give some indication of the Country's wealth as well as the intensiveness of its development during the post-war years.

An important feature of the rapid economic development in the Netherlands after World War II was the abundance of many natural resources and the existence of a highly efficient agricultural base. With space a fundamental constraint on industrial expansion, careful government planning and a series of reclamation programmes provided the impetus for the Dutch economy to exploit structural changes in agriculture (encouraged by the CAP) and become increasingly competitive within the Community's industrial sector. However, as Parker (1979) argues, the factor of most fundamental importance was (and remains) the country's geographic centrality in relation to the rest of the Community, in particular its position at the mouth of the Rhine. With Rotterdam, the main Dutch port, leading into the heart of Western Europe, the extension of intra-Community trade was a major priority for the Dutch Government during negotiations over the Common Market.

The same reasoning applied to the CAP, with the Dutch Government eager to accelerate trade liberalisation and establish further (and more profitable) markets for her agricultural exports. The proposed variable levy system was acceptable to the Dutch, but they argued against the French proposal for minimum prices. Indeed, as far as common prices were concerned, the Dutch could see nothing to be gained from basing the common level on a compromise between national price levels and preferred an autonomous decision on a 'desirable' level of support prices. The likelihood of higher Community support prices would, as far as the Dutch could see, merely lead to surpluses and problems with third countries over trade arrangements.

The chief objective of the Dutch Government was clearly to force her partners (particularly Germany) to open up their markets to Dutch food exports, particularly of dairy produce, which formed the largest proportion of Dutch agricultural production. To this end, they were willing to accept the disadvantages of higher prices and limitations on cheap third country imports of grain, a product in which the Netherlands was particularly deficient.

From the Dutch farm lobby came the most favourable reaction to the Commission's proposals on the CAP (Lindberg 1963). The main farm union, the Landbouwschap, was in favour of accelerating the adoption of the CAP, provided that all protective devices - import quotas, export aids and minimum import prices - were eliminated. The efficient Dutch farmers wanted to exploit their comparative advantage within the Community but were also seeking to maintain long-established trade links with countries outside the Community. Thus, they favoured low support price levels, particularly for grains, which were of crucial importance to the milk and livestock sectors, and criticised the Commission for being too protectionist against third countries.

The problem of cereal prices was particularly complicated in the Netherlands. As Table 3.3 illustrates, the chronic deficiency in wheat during the 1950s meant it was essential for the Government and Dutch dairy/livestock producers alike: a) to keep common prices low; and b) to maintain links with cheap grain sources outside the Community. However, given the relatively high level of cereal prices established, Dutch farmers responded by rapidly and substantially increasing wheat production, thus doubling the self-sufficiency level by 1970 and reducing the net import requirements for grain.

Table 3.3: Self-Sufficiency Levels (%) in Wheat for the 'Six' (1955/56-1966/67)

	Average for 1955/56 to 1958/59	Average for 1963/64 to 1966/67
Germany	63	79
France	105	127
Italy	103	95
Netherlands	28	57
Bel-Lux	65	72
EC(6)	89	101

Source: Raup (1970, p.153).

This is just one example of the benefits which accrued to Dutch farmers directly as a result of the adoption of the CAP. Dairy and livestock producers were to benefit even more through increased exports and higher price levels, and the Netherlands was to become one of only two net beneficiaries from the farm fund (the major one being France) during the first few years of effective common organisation of the Community's agricultural sector.

The Federal Republic of Germany

The position adopted by the German Government during negotiations on the CAP is perhaps the most interesting of all the attitudes taken by the six countries during the policy-forming years. The provision for agriculture in the Treaty of Rome was seen by many commentators (see, for example, Marsh and Ritson 1971, Clerc 1979, Swann 1981) as the result of a trade-off between France and Germany, with access to an industrial common market for the Germans to be compensated by access to an agricultural one for the French. This might suggest that Germany's interest in the CAP and agricultural problems in general was relatively unimportant. However, as any analysis of the German agricultural sector reveals, agriculture was of considerable economic and political significance during the 1950s and Germany subsequently took an active role throughout negotiations, making substantial concessions (in the interests of the 'European Movement') but ensuring some degree of continuity in the nature of support accorded to her farmers.

As Table 3.1 shows, in terms of area and employment, the German agricultural sector was substantial in the mid-1960s. However, immediately after the war the major problem facing German agriculture was inefficiency. The vast majority of farms were very small in size, over-manned with little mechanisation and thus operated at relatively low productivity levels (per hectare and per man). The land consolidation act of 1953 went some way to easing the problem, by encouraging the merger of small farm units, and between 1959 and 1971 some 780,000 farms were merged into larger, more economic holdings.

The main reasons behind Germany's relatively backward agriculture lie in the tradition of protectionism in the agricultural sector, dating back to Bismarck's tariffs on grain imports in the 1870s (Tangermann 1979a). While other West European countries chose to liberalise agricultural trade, the Germans embarked on a century of import levies and high national support prices. The proliferation of small inefficient farms put pressure on farm incomes, but, by responding to the problem through high domestic support levels, the German Government merely further inhibited the structural adjustments necessary to help the industry become competitive. This, coupled with a growing population, increased Germany's food import requirement and necessitated higher levels of border protection in order to avoid depressing domestic farm price levels. Thus, as a nation accustomed to regulating agricultural trade, yet deficient in many key products, notably grain, the German Government, when faced with the expectations of the French and (to a lesser extent) the Italians and the Dutch, discovered that its views on the form and objectives of the CAP differed substantially from those countries with considerable export potential, whose major interest was in the liberalisation and expansion of agricultural trade.

As a major food importer, Germany's interests were contrary to those shared by France, Italy and the Netherlands, but, with an inefficient agricultural sector representing some 12 per cent of the total population, the German Government also had to consider the support of its farmers' incomes. Not surprisingly, the nature of the policy which was eventually agreed reflected these contrasting interests, not only between Germany and the other member state but equally within the German economy itself.

The German farm organisations, aware of the potential threat which the CAP posed to farm income levels in Germany, took a particularly 'aggressive' stance towards the negotiations right from the start. Enjoying the highest price level in the European Community, as well as the 'benevolence' of the German Government, the Deutche Bauernverband (DBV), the main farm union, insisted upon maintaining these advantages within the Community (Lindberg 1963). It criticised the Commission's proposals for lower common support levels on the basis that consumers (in Germany if not elsewhere) enjoyed a higher standard of living than farmers. Structural policy was not considered to be necessary within the framework of the CAP, as the DBV saw this area of support as predominantly of national concern, to be operated at a domestic rather than Community level. The concept of a Common Market would only be accepted if trade liberalisation came gradually and was preceded by the removal of 'unfair' competitive practices. As far as the DBV was concerned, the problem was not the high level of German prices but the fact that costs of production elsewhere in the Community (particularly in the Netherlands) were kept artificially low, through the use of export subsidies and price controls. They could not, therefore, agree to the approximation of national prices until domestic policies and production costs had been harmonised throughout the Community. Not surprisingly, the DBV did agree with the Commission on the principle of Community Preference, a theme to which it had grown happily accustomed.

The political power of the DBV at the time of the negotiations on the CAP meant that the German Government had little room in which to manoeuvre. With the need to keep domestic farm prices high (to satisfy farm income demands) and to maintain access to (cheap) non-Community sources of food imports, it could be argued that the German Government did not really want the CAP. However, the CAP's function as the 'cornerstone' of the Community meant that the German Government, rather than overtly expressing its opposition, chose to play for time, pleading for 'understanding' and 'patience', in the knowledge that the eventual adoption of the CAP was inevitable.

Unlike elsewhere in the Community, pressure on the Government to hold its ground over agricultural policy did not only come from the farm lobby but also from industry. Germany was the largest importer of food (mainly from outside the Community) among the 'Six' and many of these imports were tied to industrial exports by various kinds of barter deals. For example, German grain purchases from Argentina were linked to Argentinian purchases of German manufactured products (Lindberg 1963). This meant German industry had an interest in maintaining trade links with third countries.

Despite the internal pressures and attempted prolonging of an agreement on the CAP, the German Government was forced to make considerable concessions during the final negotiations of 1962 and, as Neville-Rolfe (1984) suggests, these were largely granted to show that the Germans were now 'good Europeans'. By the end of the marathon Council sessions over the implementation of the CAP, Chancellor Adenauer had clearly accepted the CAP as an integral part of European unity.

It is important to remember that the Treaty of Rome was signed little more than a decade after the cessation of hostilities between Germany and the rest of Europe. In addition to the economic advantages to be gained from an extended, tariff-free Common Market, the German Government was also seeking political 'refuge' through the establishment of a truly united Europe, both economically and politically. It therefore considered that the domestic losses (both economic and political) incurred from the CAP were worth bearing in the interests of progressing towards the broader objective of European integration.

Belgium and Luxembourg

The combined arable area of these two small countries represents less than 3 per cent of the total European Community arable area and agricultural production prior to the establishment of the Community represented less than 5 per cent of the total output of the 'Six' (see Table 3.1). As the two smallest countries in the Community, Belgium and Luxembourg have traditionally lacked political and economic influence within Western Europe. Indeed, the existence (or at least independence) of Luxembourg was, until 1867, largely in the hands of its more powerful neighbours. However, with the two countries at the very heart of the Community, often acting as the driving force within the 'European Movement' and with the capital of the Common Market established at Brussels, they have enjoyed a degree of political recognition which reflects the symbolic importance of Belgium and Luxembourg as integral members of the European Community.

With both countries constrained by physical limitations, and Belgium further inhibited by a population density exceeded only by that of the Netherlands, they entered negotiations over the CAP with considerable food import requirements. As a result, facing similar problems to Germany, but lacking the industrial potential, Belgium and Luxembourg often found themselves caught between promoting the success of European integration and protesting against the financial consequences of the CAP.

If they did have an ally among the other members of the Community, then Germany was the most likely candidate. Indeed, the Belgian Boerenbond (the chief Belgian farm union) aligned itself totally with the position adopted by the German DBV, seeking to suppress 'unfair' competitive practices and extend the accession period to allow for the harmonisation of national policies.

A similar 'alliance' was established at the governmental level, with the Belgian Government closest to the German position of harmonisation before liberalisation (Lindberg 1963). Belgian agriculture was relatively backward, which meant higher prices would help to support farm incomes. But substantial import requirements suggested rather lower support price levels within the Community. In the end, the Belgians were less committed to high prices than the Germans and generally went along with the Commission's proposals for an approximation to the existing national price levels.

The position of Luxembourg more closely reflected the requirements of a relatively small economy. More attention was given to the support of individual farm incomes and the Commission was considered to be too concerned about trade with third countries. The Luxembourg Government also felt the Community needed a longer period of time to adopt common policies, and an agreement over the price level should, it believed, have been preceded by measures to reduce costs and compensate farmers for the structural adjustments which they would have to make.

Given the relative inefficiency of agriculture in both countries and their low levels of self-sufficiency in foodstuffs, it was almost inevitable that neither country would gain any net benefits from the CAP. During the 1960s Luxembourg came out as a marginal 'loser' from the farm fund, with the small net gain from structural aid insufficient to cover the inevitable deficit from the guarantee section of FEOGA.

Belgium came out of the 1960s with a much bigger 'deficit', second only to Germany, with a negative balance on both aspects of CAP support. However, on the positive side, both countries had reduced their import requirements for most food products, particularly cereals, and the Benelux alliance had itself become stronger during the period of intensive negotiations between the 'Six' on common prices and the financing of agricultural support. Thus, on balance, what Belgium and Luxembourg lost on the financial side they gained (to some extent) on the social and political side, from the progress made towards European integration and the effective redistribution of political power among the member states and the Commission.

The Commission

The importance of the Commission's role during negotiations on the CAP were twofold: first, as an intermediary between member states, it was able to help break down national 'barriers'; and second, using its (albeit limited) powers as an executive authority, it was able to influence the course of events directly on its own initiative.

In establishing the Common Market and implementing the CAP, the Commission undoubtedly succeeded in promoting the mutual benefits which were likely to accrue to member states through the process of economic and political integration. The fact that the 'Six' were able to reach a compromise on the key issues shows that in the final analysis they were willing to gamble with votes at home rather than stand accused of blocking the advance towards closer European unity (Lindberg 1963). Most commentators (see, for example, Harris *et al.* 1983, Tangermann 1980, Lindberg 1963) agree that it was the mediating and brokerage activities of the Commission which made this possible, and the fact that the final agreement closely reflected the Commission's original proposals would seem to confirm this.

Despite mutual commitment to the European Movement from the 'Six', only the Commission, acting on the Community's behalf, was able to place national arguments into the global perspective, and subsequently to coerce or encourage member state governments to make concessions (while granting safety clauses and provisions for 'special circumstances') in order that an agreement might be reached. While one might choose to criticise the Commission for its lack of precision over the stated objectives of the CAP, one cannot deny that, without its assistance over crucial areas of national conflict, any agreement would have been extremely difficult if not unlikely.

However, in its quest to accelerate the harmonisation of national policies and cement a firm foundation for the future of the EC, the Commission was perhaps guilty of underestimating the economic (and inevitably political) importance of those problems which caused most debate.

The best example of this is illustrated by the impact which the implementation of the CAP had on Community expenditure during the 1960s. As has already been stressed, nobody had anticipated the degree to which the Community's farmers would respond to the adoption of the CAP. In every country farm output expanded, surpluses (notably of sugar and butter) had already appeared by the end of the decade and, as Table 3.4 shows, the cost of the CAP had escalated beyond all expectations, with FEOGA expenditure accounting for some 95 per cent of total Community expenditure in 1969.

With this financial commitment to the CAP established, it would clearly be difficult to progress with the expansion of the Community to other areas of common organisation. The Commission, like the Governments of the 'Six', had recognised the importance of establishing the CAP, but it had not considered how the policy (and the sectors to which it related) would develop. As a result, the Commission was forced to look again at the objectives of the CAP and the means by which they were to be respected. In effect, it was obliged to take a closer look at those issues which had originally caused so many problems, not only for the creation of the Community as a whole, but particularly with the adoption of the CAP.

Table 3.4: The Growth in FEOGA Expenditure (1962/3-1968/9) (million UA)

	Agricultural Expenditure		Total Community Expenditure	Agricultural Expenditure as a % of Total Expenditure
	Guarantee	Guidance		
1962/3	24	7	32	96.8
1963/4	42	14	57	98.2
1964/5	136	45	181	100.0
1965/6	200	67	267	100.0
1966/7	308	103	412	99.7
1967/8	1,094	237	1,505	88.4
1968/9	1,677	237	2,031	94.2

Source: Adapted from De la Mahotière (1970, p145).

STRUCTURAL REFORM AND THE END OF COMMON PRICES

The implementation of the CAP required over a decade of intense negotiations, resulting in the resolution (albeit partial and to varying degrees of satisfaction) of a number of conflicting issues. One of these concerned the level of common prices. Too high a level would have upset the balance of payments situation in some member states, while too low a level would have adversely affected the farm income situation in others.

The eventual agreement ensured that common prices were initially set at a level closer to the higher price levels among the 'Six' than the lower ones. This increased the financial cost of the CAP to the net food-importing countries, but guaranteed a 'satisfactory' level of price support for the Community's farmers. Inevitably, the resolution of this fundamental problem was to lead to an even greater one - the growth of agricultural surpluses, which became the dominant influence on the development of the CAP from the early 1970s onwards.

'Agriculture 80' and the Structural Directives

By the late 1960s, it became apparent to the Commission that, due to the productive capacity of European agriculture and the incentives provided by a system of guaranteed prices without limitations on production, specific measures would have to be taken to alter the overall structure of the industry. Price cuts on their own, it appeared, were not the answer. First, farm ministers were unlikely to agree to significant price cuts, and second, there were arguments to suggest that small price cuts might result in a perverse supply response, with farmers (on aggregate) increasing production in order to maintain their incomes (Broad and Jarrett 1972). Thus, the Commission (and in particular Sicco Mansholt) took the view that production could only be controlled effectively by reducing the area of agricultural land as well as the number of people working in agriculture, and ensuring that those who remained were efficient enough to earn a living from lower support prices.

Commissioner Mansholt had warned farm ministers from the outset (at the Stresa Conference) that the combination of family farms and an open-ended price support system would lead to surplus production, inefficient agriculture and ever-increasing costs. Moreover the benefits (in the main) would go to those producers who could increase production, who (on the whole) were already better off. Thus, his plan, 'Agriculture 80', provided for '...a price policy devoted to restoring a more normal relationship between market and price levels, and radical land reform measures to bring farms up to a viable size and enable farmers to live as comfortably as possible' (Mansholt 1972). It aimed to reduce the farm population by five million between 1970 and 1980 and remove some 12.5 million hectares of land from agricultural use, thus increasing the size of farm units and improving labour and capital productivity. This was, however, to be achieved on a voluntary basis, with a variety of financial incentives to persuade farmers to retire early, take other jobs and amalgamate their holdings. These subsidies would be costly and agricultural expenditure would increase in the short term, but it was anticipated that by the mid-1970s substantial savings would be made via reduced price levels and the reduction of structural food surpluses (Commission 1968).

Not surprisingly, the plan was violently opposed by the Community's farmers and none of the member states were prepared openly and fully to support the measures proposed. With the exception of Italy and Benelux (where structural policy had been largely neglected) most of the member states had pursued their own structural policies for agriculture since before the adoption of the CAP. France had begun consolidating traditionally small holdings before World War II and Germany followed suit just after it. During the 1960s, the French Government introduced a series of measures to aid modernisation and improve access for young farmers to farm ownership. In the Netherlands, land consolidation was linked with incentives to early retirement by farmers in their fifties, and by 1967 these three countries were between them spending nearly 1.25 billion UA on structural programmes (Neville-Rolfe 1984).

Because of the political sensitivity of structural reform, the manner in which the Mansholt plan linked structural change to reduced price support levels and the increased financial commitment which the plan would entail (at least in the short run), the 'Six' were overall (with the notable exception of France) in agreement with their respective farm organisations in their opposition to the reform package.

The final agreement, outlined in 1971 and ratified under the structural directives agreement in 1972, fell short of the Commission's demands and well short of the original Mansholt plan. The compromise reform plan consisted of five basic points: a) the encouragement of farmers to leave the land and the provision of a retirement pension not less than 500 UA per year for farmers between the ages of 55 and 65; b) the introduction of farm development schemes providing low-interest loans and loan guarantees; c) the setting-up of information and advisory services; d) the encouragement of producer groups and co-operatives, to improve marketing; and e) the prevention of new land coming into agricultural production.

These in turn were watered down and condensed into the three socio-structural directives which were agreed in 1972, concerning: a) the modernisation of farms; b) encouraging the cessation of farming and the re-allocation of utilised agricultural area for the purposes of structural improvement; and c) the provision of socio-economic guidance for and the acquisition of occupational skills by persons engaged in agriculture.

The year in which the agreement on the structural directives was reached was Sicco Mansholt's last as Agricultural Commissioner, but, while his plan for structural reform stands out as the highlight of internal pressure for concentration of the CAP on structural change, the initiatives which were eventually agreed confirmed that price policies would remain at the core of agricultural support within the Community. While the farm ministers of the 'Six' recognised the need for the reform of the Community's agricultural structure, they were reluctant to accept the political consequences of agreeing to the appropriate reform of its agricultural policy. As Mansholt (1972) himself protested, the price policy was based on consensus politics rather than economic rationale and the failure of the 'Six' to respond to the challenge of structural reform was a 'shirking of responsibility'.

'Green Money' and MCAs - a New Concept of Common Prices

In 1962, following its successful use in the ECSC, the Commission had decided to use the Unit of Account (UA) for the purpose of fixing common support prices under the CAP. This was simply a standard of measurement the value of which was defined by reference to a weight of 0.88867088 grammes of gold, corresponding to the gold parity of the dollar at that time, as declared by the International Monetary Fund (IMF). In effect one UA was made equal to one dollar. Since each of the national currencies in the Community had a declared parity within the IMF, as arranged under the

Bretton Woods agreement in 1944, common prices fixed in UAs could be directly converted into national currencies, and, in whichever national currency they were expressed, they would be internally consistent.

The period over which the CAP was established was one of relative currency stability, with fixed (though in principle adjustable) exchange rates operating throughout Western Europe. Thus, few of the Community's policy-makers envisaged the problems which emerged in the early 1970s following the effective breakdown of the Gold Standard and the floating of individual currencies against the dollar.

Only two years after the establishment of common prices within the CAP, parity changes were implemented in France and Germany: in August 1969 the French franc was devalued by 11.11 per cent and in October of the same year the Deutschmark was revalued by 9.29 per cent. These parity changes marked the end of the only period when common support prices for products under the CAP applied throughout the Community.

The devaluation of the French franc was the first parity change among the 'Six' for eight years. Largely a result of the rising inflation, following the riots and subsequent wage increases of 1968, the devaluation was against Community regulations and, with no prior consultation sought with the Commission, it took everyone by surprise. Agricultural and finance ministers met to discuss the impending problems of trade distortion resulting from the devaluation, but ruled out any offsetting change in the value of the UA. Instead, it was agreed that French farm prices would be aligned gradually to the new exchange rate, and that for a *limited* period (not later than the end of the 1970/71 marketing year) the French Government should be obliged to place subsidies on intra-EC food imports and levy taxes on intra-EC exports to remove the 'unfair' price advantage. Thus, the Monetary Compensatory Amount (MCA) was born and, like so many of the 'temporary' measures envisaged by the Commission, it was to become a permanent feature.

The German Government's decision to revalue the Deutschmark, following months of upward pressure, was preceded by consultation and agreement from the Commission. However, its decision to impose an import levy of 11 per cent on all CAP products, to compensate German farmers for the downward effect on their prices, was not. Instead, the Commission requested a total ban on agricultural imports and this for three reasons: first, the Commission wanted to highlight the undesirability of fluctuating exchange rates in the context of the CAP; second, if a currency had to be floated it should be done for a day or two only, as it would not be able to last without importing foodstuffs for a longer period, and third, the Commission did not want the imposition of MCAs to become a permanent feature within the CAP, as this threatened the principle of the single market. Although upheld by the European Court, the Commission's view was not enforced and it was eventually agreed that a positive MCA should be applied to intervention products only and that it should be revised weekly in the light of currency movements.

Following the general election in Germany in 1970, the incoming Minister for Agriculture, Josef Ertl, agreed to the removal of the MCA on the condition that German farmers would receive compensation from the resultant decline in farm prices. This amounted to a reduction in the farmers' liability for VAT payments, to be spread over three years. The Commission defended the compensatory measures on the grounds that it would have been politically unacceptable for German farmers to face price cuts without compensations, and the alternative of maintaining MCAs would have encouraged the proliferation of MCAs elsewhere in the Community.

However, the reluctance of the French (fearful of the inflationary consequences of raising food prices) and the Germans (wary of the political consequences of lowering prices to farmers) to allow farm prices to adjust following the divergence in market exchange rates meant that the Commission was obliged to allow agricultural support prices, in national currencies, to diverge.

This resulted in the introduction of exchange rates for an agricultural unit of account (AUA) into national currencies, which differed from the market exchange rates and subsequently became known as representative or 'green' rates of exchange. If the common organisation of agricultural markets was 'temporarily' threatened by the introduction of MCAs, the creation of 'green' money effectively returned part of the control of agricultural prices to national governments.

The end of dollar convertibility in August 1971, and the 'Smithsonian Agreement' in December of that year, meant that by 1972 'green' rates were introduced for all member states as they each chose to maintain domestic price stability rather than respect common prices when exchange rates changed.

Prior to August 1971, the UA and the dollar shared the same gold parity. When the dollar was removed from the Gold Standard the Commission decided to maintain the UA/dollar link rather than that with gold. This meant that the MCA for any member state depended on the country's currency movement against the dollar.

The 'Smithsonian Agreement' was designed to restrict the degree to which exchange rates between the signatories (all the major Western trading countries) could fluctuate, in the hope of restoring stability to the currency markets. Soon after this agreement, the Community adopted a more restrictive system for itself - the 'snake in the tunnel', with a maximum margin of fluctuation of 2.25 per cent between those market rates of member states' currencies showing the greatest appreciation and greatest depreciation.

It was also agreed that the cost of financing MCAs (provision of export subsidies and so on) should be transferred from member states to FEOGA, but that they should be phased out over a three-year period, with no compensation for revaluing countries.

With the floating of the 'joint-float' currencies - those of Germany, Denmark, Benelux, the Netherlands and (periodically) France, in 1973, the UA was effectively re-defined once more, with its value linked to the 'joint-float' rather than the dollar, and fixed in terms of each of the 'joint-float' currencies. Those countries within the 'joint-float' were granted 'fixed' MCAs which only changed when the currency or the 'green' rate was revalued or devalued. Those countries outside the 'joint-float' (initially Italy and, following accession, Ireland and the United Kingdom) had 'variable' MCAs as their currencies could fluctuate more widely against the 'joint-float' currencies and hence the 'joint-float' UA.

The development of problems resulting from this new method of defining common prices and converting them into national currencies is examined in more detail later in this chapter. However, it is important to note here the extent to which this system allowed CAP support prices in the different countries to diverge, with the gap between the highest and lowest prices often greater than that which existed prior to price harmonisation in 1967 (Harris *et al.* 1983). Moreover, while MCAs did enable the CAP to continue to operate despite the breakdown in the unity of prices expressed in national currencies, they became an increasingly costly item of expenditure under FEOGA (400 million UA or 9 per cent of FEOGA expenditure in 1975 and 860 million UA or 12 per cent of FEOGA expenditure in 1977) and their proposed removal created a close link between adjustments to exchange rates and changes in institutional price levels - a link which has grown in significance from the first day they were applied.

THE SIX BECOME NINE - POLICY IMPLICATIONS OF THE FIRST ENLARGEMENT

Despite the problems of monetary instability and the growth of agricultural expenditure within the Community, the potential significance of the European Community as an economic and political body of equal power to the US and the USSR resulted in a growing interest among West European non-member countries to become more directly involved and to share the apparent benefits to be gained from internal trade liberalisation within the protective common tariff barrier. Such interest was reflected in Britain's repeated attempts to gain favour with the 'Six' during the 1960s. British efforts to join the Community failed on two occasions, first in 1962 and again in 1967, but in the accession negotiations of 1970 she was joined by Ireland, Denmark and Norway. By 1973 the terms of entry were agreed, with only Norway rejecting membership at the last hurdle, following a referendum in 1972, and on January 1, 1973, the 'Six' became 'Nine'.

The enlargement promised to have significant economic consequences for the Community as a whole, particularly as Britain provided a substantial export market for the Community's agricultural surpluses. However, while Denmark and Ireland were both heavily dependent on the agricultural sector as an employer and an earner of foreign exchange, Britain, like Germany, was much less interested in the agricultural aspects of accession (apart from the financial burden of the CAP) and was more concerned with exploiting the liberalisation of industrial markets within the European Community. Thus, while for Ireland and Denmark membership was likely to result in substantial (net) benefits accruing to their economies (via the CAP), the immediate prospects for the United Kingdom, a major food importer accustomed to relatively low food prices, were less favourable. Indeed, the implications of the CAP for the British economy had been the major factor inhibiting the commitment of successive British Governments to the European Community.

Britain's initial departure from the 'Six', during the early negotiations at the Messina Conference, was founded on: a) its overwhelming scepticism over the ability of the 'Six' to achieve anything of either economic or political note; b) its reluctance to surrender national sovereignty; c) the belief that its own economic strength and industrial competitiveness at the heart of the world economy could be maintained, a contributory factor stressed by Edward Heath (1982), the Prime Minister who eventually took Britain into Europe, in 1973.

The implications for Britain of adopting the CAP were twofold: first, the implementation of tariffs and a higher level of common support prices would increase the cost of food to the British consumer and ultimately have an inflationary effect on the United Kingdom economy; and second, as a substantial food importer (though maintaining concessionary agreements with the Commonwealth) the United Kingdom would also 'lose out' through the higher cost of food imports and the budgetary contributions to the Community, with most of the 'benefits' going to the agricultural exporting countries of the 'Nine' (Marsh and Ritson 1971). It would be an over-simplification to suggest that the United Kingdom Government accepted this prospect out of hand - the British insisted on a transitional period of at least five years (to ease the process of adjustment), gained considerable concessions for her Commonwealth partners (notably New Zealand) and forced an agreement on a level of budgetary contributions lower than that proposed by the Commission (Swann 1981). But many of the fundamental differences of opinion (in particular those relating to agriculture and the budgetary implications of the CAP) were swept aside (albeit temporarily), in the effort to secure an agreement.

While British entry clearly presented a number of potential pitfalls for the future of the Community, it also provided a potential short-term solution to the growing problem of surplus food production among the 'Six':

with 40 per cent of temperate food requirements in the UK being imported, the British market could have absorbed all the European Community surplus of butter, cheese, grain and sugar, but at a per unit cost twice that which it faced prior to accession (De la Mahotière 1970). The fact that this potential was not fully realised was due, in part, to the maintenance of New Zealand butter supplies, the agreements established with the African, Caribbean and Pacific (ACP) countries over sugar exports to the European Community, the continued UK preference for North American grain and expansion of Community production, which offset any increased demand resulting from enlargement.

In the United Kingdom, from their original hostility towards UK membership of the European Community, the opinion of the farm lobby swung around to its total support of accession by 1973. The deficiency payments system had, until the mid-1960s, worked favourably for the farming population as a whole. But at a time when British farmers were being encouraged to expand production 'selectively' (due to the rising exchequer cost of food subsidies) Community farmers appeared to enjoy more security in high guaranteed prices and an open-ended market.

Prior to accession, British farmers believed that producer prices would increase as a result of the CAP and, given the relatively efficient structure of British agriculture, that they would be well placed to compete. However, as Marsh (1979) argues, the Government's desire to maintain a relatively high 'green' rate (necessitating a high negative MCA), in order to protect consumer prices, threatened to deprive British farmers of many of the expected benefits. As it was, high world prices raised UK prices substantially, which largely offset the efforts to keep prices down (see Chapter 4).

During the first few years of membership, any potential criticism of the CAP from the British was dampened by the crisis on the world raw materials and agricultural markets. Inclusion in the European Community kept prices relatively stable and for some products (notably sugar and cereals) below world market levels. This reduced the direct burden of the CAP, helped control the rising level of inflation in the United Kingdom and, as Tangermann (1980) argues, played a significant part in the 'positive' outcome of the British referendum on continued European Community membership, in 1975.

However, this period of comparative harmony over agricultural policy was short-lived. In 1977 the Consumers' Association published a study calling for reduction in agricultural prices (particularly of surplus products), the National Farmers' Union (NFU) began to voice its discontent over the maintenance of high negative MCAs, and the notion of the *'juste retour'* became a prominent argument in favour of CAP reform.

In contrast to the economic and political significance of Britain's accession to the European Community, Ireland, like the other political lightweights in the Community (that is, Belgium and Luxembourg), could expect to do little itself to influence policy developments. However, in contrast to its small country allies, Ireland, with its agriculture becoming increasingly competitive, was able to sit back and reap the rewards from the decisions influenced most heavily by the more powerful member states. As long as agriculture remained the major beneficiary from the Community budget, the Irish Government had no reason to complain. Not only has Ireland maintained a modest but steady agricultural trade surplus among the 'Nine', it is also the second largest net beneficiary from the Community budget.

The participation of Denmark in the European Community and the process of economic and political integration is characterised by Haagerup and Thune (1983) as being largely non-committal. Sharing many geographical similarities with the Netherlands, a country of about the same size, low-lying with maritime traditions, an intensive agricultural sector and for the most part prosperous in relation to her European neighbours, Denmark's main contrasting feature is her reluctance to become involved with Community affairs. Such a stance towards the Community is in many ways ambiguous: while the Danish economy, particularly the agricultural sector, clearly stood to gain from European Community accession, there was a marked reluctance, particularly among the left-wing political parties, towards this commitment. This was partly due to the difficulty of reconciling Community membership with a long-standing association with the other Nordic countries and partly the result of Denmark's dependency on the United Kingdom as a trading partner.

The Danish economy is highly dependent on international trade - its exports being worth around one quarter of GDP - and the agricultural sector in particular enjoys a substantial trade surplus. With the United Kingdom being her main trading partner (accounting for around 20 per cent of total exports and about 50 per cent of agricultural exports) Denmark was not in a position to contemplate joining the European Community from the outset, but instead was 'obliged' to follow Britain in joining the European Free Trade Area (EFTA) in 1959, along with the other Scandinavian countries. Similarly, it was not until the early 1970s, when the UK was clearly set on joining the Community, that the Danish Government adopted a similar policy and entered into negotiations with the 'Six'.

The importance of agriculture in the Danish economy was a major factor in favour of Danish accession: up to 25 per cent of the working population was involved in agriculture during the fifties, but a series of government development programmes after World War II helped to rationalise Danish agriculture and render it one of the most efficient in Europe. This was achieved by exploiting the 'gaps' in the agricultural markets of her trading partners, in particular Britain and Germany, and replacing cereal production with pastoral produce, aided by a policy of favourable feed grain prices.

However, as Tangermann (1980) points out, the stimulus for Community membership resulted from the decline in the growth of agricultural production, food exports, farm incomes and agricultural investment during the 1960s, due to the increasing levels of protection and high rates of agricultural expansion in other countries. In this situation, European Community entry and the associated unrestricted access to the traditional markets of Britain and Germany were seen to be extremely attractive. Indeed, the relative economic success of the Danish economy as a whole, and the agricultural sector in particular, under the Community umbrella, has resulted in the Danish Minister of Agriculture adopting a cautious attitude towards price increases in the Council of Ministers. This is not out of a desire to reduce the benefits accruing to Danish farmers from the Community budget, but rather a pragmatic approach to ensuring the maintenance of the CAP in (more or less) its current form.

NATIONALISM AND INSTITUTIONAL INITIATIVES

As anticipated, the first enlargement of the European Community in 1973 marked the beginning of a new era for the development of the Community in general and the consideration of the CAP in particular. The accession of Britain posed a significant threat to the CAP in its present form, but, more important, British attitudes towards the function of the European Community and the role of member states within the Community framework encouraged each member state to look more closely at the budgetary transfers and resource flows which resulted from the adoption of common policies, jointly financed and under the corporate management of the Commission.

The results of this growth in nationalism within the Community, directly related to the financial implications of the CAP, were twofold. First, the 'pro-CAP' member states (notably France, Ireland, the Netherlands and Denmark) sought to protect the national benefits accruing from the CAP, by pushing for higher support prices and maintaining the production of surplus commodities (particularly cereals and milk), which in turn exacerbated the budgetary problems of the CAP. Second, those member states whose interests remained in CAP reform (in particular Britain and Germany) pushed for a prudent price policy with direct compensation for the perceived 'unfair' distribution of the costs of the CAP. For the United Kingdom, this took the form of budgetary rebates, while, for Germany, the Commission took a lenient position over the removal of positive MCAs.

Such polarisation among the member states over the operation of the CAP made any rational discussion of policy reform at best difficult and, more often than not, impossible. Thus, it was left to the Commission to seek a solution to the conflicting problems facing the 'Nine' during the 1970s.

However, while agricultural expenditure grew, along with surplus production, the Commission provided few alternatives in a series of reform documents which served merely to reiterate the basic objectives of the CAP, increase the financial pressure on the Community budget and intensify the conflicting interests between member states.

Features of Nationalism Within the CAP

Following the implementation of the basic mechanisms of the CAP and the establishment of support price levels, decisions on policy developments were to be taken by majority voting in the Council of Ministers. However, the 'Luxembourg Compromise' effectively established the power of veto over decisions of 'vital national interest' and, if the 'Six' found this objective undesirable, the possibility of implementing a majority voting system with the 'Nine' was remote. Thus, as Marsh and Swanney (1980) conclude, while the European Community was founded on the principle of a pooling of sovereignty over Community decisions, in practice CAP-related issues have been regarded as being of such vital interest that the unanimity rule has been maintained, with the result that major decisions require a compromise of national interests rather than an expression of common interest.

Two aspects of the preoccupation with national interests are examined here: first, the manner in which the economic implications of the CAP are expressed between the member states - income transfers from net importing countries to net exporting countries; and second, the range of stances adopted by member states, in defence of national interests (the maintenance of benefits or the reduction of costs), within the corporate framework of the Community's decision-making process.

Ritson (1979, 1983) points out the significance of the defence of national interests within the Community in relation to the economic analysis of the CAP. As far as member states are concerned, interests lie not in the overall performance of the CAP in attaining basic objectives (as defined in the Treaty of Rome), but in the impact of the CAP on financial transfers between member states (see Buckwell *et al.* 1982). The importance of these transfers is in turn largely a function of the importance of the agricultural sector in each member state, or, more explicitly, the political influence of the farm lobby. Some indication of the relative political significance of agriculture in the 'Nine' is given in Table 3.5, which suggests that agricultural policy should be highest on the political agenda in France, Ireland and Italy and lowest in the United Kingdom. However, national costs of the common financing of the CAP diverge between member states and collective price decisions influence individual member states unevenly, with agricultural exporting countries gaining (on the whole) at the expense of agricultural importing countries.

Table 3.6 is one illustration of the extent to which the burdens of common price increases have been distributed among the nine (for alternative measures see Koester 1977 and Buckwell *et al.* 1982). The United Kingdom, Germany and Italy (the major food importers) suffer the highest welfare losses as a result of the distribution of budgetary contributions. This perception of 'inequality' becomes all the more striking when one considers that Britain and Italy, two of the major contributors to the Community budget, also rank among the poorest members of the European Community (see Table 3.5).

While it is not the objective here to justify or criticise the attitudes adopted by member states during the inflationary period of the 1970s, it is evident that the problems of an unequal distribution of costs and benefits emanating from the CAP were of crucial importance as far as the attitudes of member states towards policy changes were concerned. However, rather than concentrating on the longer-term rationalisation of the CAP, there was (and remains) an overriding tendency to seek the resolution of national conflict via the central policy instrument, the support price.

Table 3.5: Indicators of Political and Economic Weight, EC(9)

	Germany	France	Neth.	Denmark	Italy	Ireland	UK	Bel/Lux
GNP/Head ($) (1976)								
	7510	6730	6650	7690	3220	2620	4180	7020
Average Farm Size (ha) (1976)								
	14.0	25.0	14.7	23.1	7.5	20.56	4.7	14.2
Agricultural Population as a % of Total (1979)								
	6.2	8.9	6.0	8.3	14.9	21.00	2.6	3.3
Salaried Agricultural Labour as % of Total (1979)								
	17.0	20.0	25.0	25.0	37.0	11.00	58.0	10.0
National Expenditure on Agriculture (1977)*								
	13.8	23.7	4.8	9.2	7.5	12.90	21.3	9.3

*: As a proportion of the value of final agricultural production.
Source: Ritson and Tangermann (1979, p126); Harvey (1982, p183).

It is with regard to the positions adopted by member states over support price levels that the importance of the agri-monetary system is again highlighted. The very fact that national attitudes towards agricultural policy differed among the 'Nine', along with the perceived role of the CAP in regulating farm prices, meant that MCAs served a vital function during the 1970s in allowing member states a degree of freedom over domestic farm prices, when the imposition of common prices (and the resultant increase in income transfers) would have increased the tension between member states as well as the political strain on the Community as a whole. Moreover, the use of MCAs returned some sovereignty over agricultural policy to member state governments, a factor which, as Ritson and Tangermann (1979) illustrate, resulted in most member states pursuing national policy objectives (at least with regard to price support) within the framework of the CAP.

Table 3.6: Costs and Benefits of the CAP with Respect to Common Price Changes

	Ratio of Total Welfare Loss to Total Gains from 10% Price Increase	Average Cost per Unit Increase in Farm Incomes			
		FEOGA Expend.	VAT Contr.	User Welfare	Economic Welfare
EC(9)	1.79	0.56	0.76	0.89	0.66
Germany	2.20	0.39	0.96	1.09	1.05
France	1.61	0.64	0.72	0.77	0.49
Italy	1.98	0.73	0.65	1.25	0.89
Netherlands	1.05	0.62	0.50	0.47	-0.03
Bel/Lux	1.78	0.33	0.66	0.69	0.35
United Kingdom	2.51	0.31	1.12	1.21	1.34
Ireland	0.51	0.90	0.21	0.31	-0.48
Denmark	0.79	0.99	0.37	0.30	-0.34

Source: Harvey (1982, p 178).

Member states have, in practice, the right to maintain MCAs rather than adjust their 'green' rates if they wish. Thus Britain was able to maintain a (relatively) cheap food policy within the CAP by refusing to devalue the 'green' pound in line with the market rate, thereby sustaining large negative MCAs. A similar position was adopted by Italy, a large importer of 'Northern' products, which progressively sought to satisfy the Italian farm lobby via structural measures and direct aid from the Community budget. The ambiguity of Germany's maintenance of relatively large positive MCAs, as a net food importer, reflects the importance of the farm vote in Germany and the relative prosperity of Germany among the 'Nine' (see Table 3.5). Denmark, as noted in the previous section, adopted a policy of directly linking the 'green' krone with the market rate, in order to satisfy domestic farmers and exploit the benefits of jointly financed export refunds on high-priced CAP products. A similar (low MCA) policy was adopted by Ireland and the Netherlands and, among the 'pro-CAP' countries, only France chose to maintain high negative MCAs, although, as Ritson and Tangermann (1979) point out, the use of MCAs in France during this period more closely reflected the use for which they were originally designed, to aid the adjustment of the French economy to successive devaluations of the franc.

While the proliferation of MCAs undoubtedly smoothed the path for the Community over the issue of CAP reform, it also staved off the day when the basic operation and underlying principles of the CAP would be reviewed. As Tangermann (1979b) points out, the overall tendency of member states to push the CAP (or at least the level of CAP support prices) as far as possible in the direction of the policies they would have pursued in its absence, coupled with the inexorable increase in common prices following the link between the AUA and the strongest 'joint-float' currency, the Deutschmark, served to exacerbate the problems of surplus production and the level and distribution of agricultural expenditure.

The introduction of the European Monetary System (EMS), in 1979, introduced an important stabilising factor into European currency markets and member states' exchange rates. An extension of the 'snake' system , based on a weighted currency 'basket', with the European Currency Unit (ECU) replacing the Unit of Account, the EMS was introduced in an attempt to stabilise exchange rates following the collapse of the dollar under President Carter's policy of 'benign neglect' (Fearne and Perry 1988). Aimed at establishing greater monetary co-operation, the EMS went some way to halting the growth of negative MCAs in the weak-currency countries, and reduced the potential for support prices (in national currencies) to diverge between member states.

Nevertheless, the Commission had effectively lost much of its managerial control over the implementation of the CAP, and the resolution of conflicting interests became increasingly difficult as national attitudes became further entrenched. However, rather than responding to the challenge, as the Community's policy formulator, of seeking a resolution not only to the conflicting interests of member states but to the growing problems facing the Community's agricultural sector as a whole, the Commission published a series of 'initiatives' which reiterated the basic objectives of the CAP, the economic irrationality upon which they were based and the political tension which they had led to, but offered little in the way of possible solutions.

The Commission Dilemma Over CAP Reform

Since accession in 1973, the United Kingdom has been at the head of those countries (albeit a conspicuous minority) in favour of the wholesale reform of the CAP. In recognition of Britain's stance towards agricultural policy and in an early attempt to address the problems of market imbalance and excessive expenditure in the agricultural sector, the Commission produced the first (following the Mansholt Plan) of periodic reviews of the policy entitled *Improvements of the Common Agricultural Policy* (Commission 1973). Part of the blame for the problems resulting from the CAP was placed on the lack of progress in other areas (notably monetary union and regional policy), but the Commission recognised the need to reduce the disequilibria in certain markets and cut back the level of guarantee spending. Ironically, the one thing that was keeping the Common Market functioning, the MCA system, was considered a threat to market unity and the Commission wanted its removal by 1977.

There was little in the way of new ideas presented in the document and in any case the surge in world commodity prices around this time took some of the pressure off CAP reform. However, in 1974, following the request of the newly elected British (Labour) Government to renegotiate the terms of entry, the Commission was asked to prepare a *Stocktaking of the Common Agricultural Policy* (Commission 1975), which attempted more firmly than before to comment on the CAP relative to the objectives laid down by the Treaty of Rome. A careful reiteration of the five basic principles of the CAP was supported by evidence to show how the CAP had (more or less) stuck to the these objectives. However, four main problems were noted: a) the failure of price policy to reflect the market situation; b) the failure of structural policy to increase productivity and reduce regional disparity; c) the continued threat to market unity posed by the MCA system; and d) the growth in budgetary expenditure under the CAP.

Once again the problems had been noted but no solutions were offered and, despite Britain's preoccupation with the CAP, agriculture was not a key issue in the final renegotiation terms agreed in Dublin in 1975. High world prices for cereals, sugar and animal feed protein made the CAP appear less protectionist and again blunted criticism, condemning the issues raised by the Commission to scant recognition within the Community (Harris *et al.* 1983).

It was not until 1979, when reform proposals were triggered by the dangers resulting from CAP expenditure growing faster than 'Own Resources', that the Commission produced a further communique - *Changes in the Common Agricultural Policy to Help Balance the Markets and Streamline Expenditure* (Commission 1979), which effectively marked the beginning of a number of serious attempts to modify the CAP in order to counter the growth of structural surpluses and save it from the pressure to reduce expenditure. Thus, two main proposals were made: a) to move towards closer market balance, especially for milk and sugar; and b) to enforce producer participation in the cost of surplus disposal.

What followed was a number of Commission documents concentrating explicitly on the method and nature of reform, which served to encourage a concerted effort among the Community's policy-makers to consider an alternative method of agricultural support in the European Community. The reform debate of the 1980s is discussed in Chapter 15, but it is important to note here that it was not the initiatives of the Commission which ultimately provoked the consideration of policy reform by the member states, but the real and 'tangible' threat of bankruptcy as a result of unabated CAP expenditure.

The significance of the documents mentioned above and the institutional pressures for reform during the 1970s was essentially twofold. On the one hand, all of the Commission's early reform documents recognised the budgetary and market problems resulting from the CAP and stressed the need to reduce market imbalances and cut back on agricultural expenditure. On the other, the Commission took great pains to reiterate the goals of the CAP and its support of price guarantees as the central policy. Thus, the initiatives emanating from the Commission, during this period, reflected an (inevitable) ambivalence in attempting to accommodate a diverse range of national perspectives towards agricultural policy and the dichotomy of seeking reform without changing the basic structure of agricultural support accorded to the 'Nine'.

THE NINE BECOME TWELVE - AGRICULTURAL PROBLEMS OF MEDITERRANEAN ENLARGEMENT

On January 1, 1986, five years after the official entry of Greece into the European Community and two years after the end of the Greek accession period, Spain and Portugal became full members of the Community. Unlike negotiations with Greece, neither party to the most recent Treaty of Accession was particularly anxious to reach an agreement, with the Treaty relating to Spain and Portugal being finally agreed eight years after the start of negotiations. Similarly, while for most products Greece fully adopted the CAP within five years, the transition period for Spain and Portugal was extended to ten years.

This contrast in negotiations over the entry of the 'Three' is partly explained by the particular political circumstances dominating Greece's attitude towards European integration and also by the fact that, while Greek accession was not expected to rock the Community boat too much, the further expansion of the Community towards the Mediterranean, in particular to include the greater political and economic power of Spain, threatens significantly to alter the course of European integration in the future. This factor alone necessitated the more careful consideration of the implications of accession, as well as a more structured and gradual process of integration of the Spanish and Portuguese economies within the Community.

Greece

Greek involvement with the Community began as early as 1962, with the Association Agreement, which established favourable arrangements (reduced levies, increased import quotas, and so on) for Greek exports to the

Community. However, it was in 1972, following the Paris Summit, when the 'Six' agreed a global Mediterranean policy, that the increased involvement of Mediterranean countries was provided for, with a view to creating a free trade area in industrial goods between the Community and each of the Mediterranean countries, with reciprocal tariff reductions, yet maintaining Community preference for agricultural products, with only limited *mutual* concessions.

Following the events of July 1974 in Cyprus which led to the fall of the Greek military Government, the new (civilian) Greek Government, headed by Constantine Karamanlis, declared its firm intention of joining the European Community, on the grounds that accession would help the Greek economy, consolidate the country's democracy and (arguably most important at the time) strengthen its defences and its position against Turkey (Tsakaloyannis 1980).

Negotiations on Greek accession began in the mid-1970s, with the Greek Government prepared to take the economic risks of speeding up the transitional period in order to profit from the political security which membership of the Community would, it hoped, ensure. From 1975, Greek exports to the Community were exempt from tariffs (apart from compensatory taxes on agricultural products), as were two-thirds of exports from the Community to Greece (the other third accounting for 56 per cent of customs duties), and negotiations were enthusiastically pursued during the late 1970s, with the Greek Government making substantial economic concessions (Tsakaloyannis 1983). Full membership was finally achieved in 1981, with the alignment of prices, the reduction of duties and the application of the common external tariff established just three years later, in 1984.

The apparent lack of enthusiasm of the 'Nine' towards Greek entry into the Community was largely due to the potential threat which the Mediterranean countries posed for the future of the CAP and the development of agricultural expenditure under the Community's financial umbrella. Table 3.7 illustrates the relatively important role which agriculture plays in the Greek economy, with over a quarter of the working population involved in agriculture and 14 per cent of GDP generated within the agricultural sector. Moreover, agricultural exports constitute 29 per cent of the total value of Greek exports and, as a net food exporter, Greece stands to gain from the CAP. However, production costs (even in the most efficient areas) are higher in Greece than in most of the member states and the high labour/land ratio reflects the proliferation of small holdings, where the limited ability to respond to high support levels is unlikely to significantly effect overall self-sufficiency of foodstuffs in the European Community as a whole (Siotis 1983).

Perhaps the biggest fears within the Community over the impact of Greek accession were voiced by the French and Italian Governments, who anticipated Greek competition over particular products, such as fruit and vegetables. This was a fast growing sector during the 1970s, with its share in total agricultural production rising from 15 per cent in 1972 to 25 per cent in 1978. However, while labour costs may be lower in Greece than elsewhere in the Community, the level of production per worker is also lower, making labour costs per unit of output comparable to those in the rest of the European Community (Pepelasis 1983).

Table 3.7: The Importance of Agriculture in Greece, Spain and Portugal (1987)

	Greece	Spain	Portugal
Percentage of total employment in agriculture	27.0	15.1	22.2
Arable area as a percentage of the EC(12)	4.3	22.9	4.3
Agricultural population as a percentage of the EC(12)	9.8	17.4	9.4
Agricultural exports as a percentage of total exports	28.6	16.9	12.9
Agricultural imports as a percentage of total imports	15.2	21.2	30.5
Agriculture as a percentage of GDP (1986)	14.4	5.6	6.5

Source: Commission (1988).

While Greek accession added 7 per cent to the total utilised area of the Community and 13 per cent to the labour force, it only contributed a further 4 per cent to the value of agricultural production. The relatively small size of Greek agriculture and the degree of complementarity between agricultural production in Greece and that in the rest of the Community, as well as the lack of political power within the Council of Ministers, leave Greece playing a relatively small role, individually, in the decision-making process of the Community. However, as one of a growing number of Mediterranean countries incorporated within the European Community, the *collective* action of Greece, Spain and Portugal is beginning to have a significant impact, by shifting the geographical and political 'balance' of Community policies.

Portugal

Portuguese involvement in the 'European Movement' is a relatively recent phenomenon, with the African colonies generally regarded, up to the 1950s, as a sufficient outlet for future economic expansion (Cravinho 1983). However, the first step towards economic integration with the rest of

Europe came with Portugal's membership of EFTA (paradoxically juxtaposed to the EC), which if nothing else highlighted the importance of European markets to the Portuguese economy: EFTA's share of Portuguese export markets increased from 18 per cent in 1958 to 35 per cent in 1972, and, with the importance of colonial markets declining from around 25 per cent to 19 per cent over the same period, the ruling elite was forced to accept the inevitability of increased involvement with the EC.

This became even more necessary after the first enlargement of the Community, which took Portugal's biggest client, Britain, out of EFTA and into the European Community. Prior to British accession, the United Kingdom represented 52 per cent of Portugal's EFTA markets and along with Denmark over 21 per cent of total Portuguese trade. Thus, overnight, Portugal had to adjust her trading pattern or risk a dramatic loss of export earnings.

The initial result was the Association Agreement, signed in 1972, which enabled Portugal to maintain trade with the former EFTA members, at preferential tariff rates, progressively over a five-year period, to 1977, when negotiations on full membership began.

The 'Nine' played an important role in facilitating Portuguese involvement in the Community by offering financial aid during the mid-1970s, to help avoid the eruption of civil war and to support the pluralistic democracy (Cravinho 1983). The first emergency loan of 180 million ECU was granted in 1975, and, as a result of the increased co-operation between Portugal and its European neighbours, the first constitutional (socialist) Government announced in its programme, in 1976, the intention to begin negotiations on European Community accession immediately.

With the exception of the Communists, most of the Portuguese political parties were in favour of accession, largely due to the potential economic benefits which were expected to accrue to the Community's poorest member state. The Commission was less concerned with the economic aspects of Portuguese accession, but rather more interested in aiding the democratisation process, although the development of a global Mediterranean policy was facilitated by the inclusion of Portugal, along with Greece and Spain, *within* the Community boundary.

A major factor contributing to Portugal's state of relative poverty is her dependence on a largely stagnant agricultural sector. As in other Mediterranean countries, climatic conditions have contributed to poor agricultural performance, but, more significantly, low productivity levels (of both land and labour), a lack of entrepreneurial talent and poor cropping systems have 'transformed Portuguese agriculture from a dynamic sector into a static pool of resources' (De Abreu 1983).

Production growth rates were more or less static during the 1970s, with the notable exception of the fruit sector, which grew at an average annual rate of 6 per cent. Yields of most crops, particularly cereals, have similarly stayed more or less constant, while the arable area has declined, largely due to out-migration and the limitations imposed by increased mechanisation. Traditional crop rotations have failed to meet the pattern of demand, and the level of overall food production has steadily declined from 1960, turning a small food surplus into a significant food trade deficit of over one billion ECU in 1987, with agricultural imports representing 30 per cent of the total value of imports (Commission 1988).

Despite limited land reform, aimed at encouraging the growth of co-operatives and self-managed farm units, the more recent development of part-time farming (particularly in areas of industrialisation) and the lack of financial aid and co-operation between the Government and the farm organisations has left the structure of Portuguese agriculture dictated by very small units - of the 350,000 holdings, over 75 per cent have less than 5 hectares.

It is evident that Portugal is faced with a number of problems following accession which have many similarities with those experienced by Greece. The agricultural sectors of the two countries have much in common, with low productivity levels, a proliferation of very small units, and concentration on the production of traditional Mediterranean products. However, as with Greece, the agricultural picture is not all gloomy. There is an element of complementarity between the pattern of production in Portugal and the 'Nine', with potential for exports in wine, nuts, tomatoes and some fruits.

Thus, as a political lightweight, of little economic significance, Portugal, like its Mediterranean neighbour Greece, is unlikely to have a significant impact on the nature of Community policies or the decision-making process in the future. However, the inclusion of Portuguese agriculture in the CAP will increase the need for structural reform and (if Portugal is to 'survive' in the Community) necessitate a broader 'Communautaire' attitude amongst the 'Nine', with the recognition of the need to channel considerable financial aid, through the Community budget, to the poorer areas of the Community.

Spain

With Greece and Portugal the two smallest members of the 'Twelve' (with the exception of Luxembourg), the Spanish economy is by far the strongest of the 'Three', being the fifth largest in the Community, with significant export potential in certain agricultural products (notably fruit and vegetables, wine and olive oil) despite being a net food importer.

Jordan (1985) highlights the extent to which agricultural factors contributed to the difficulty of negotiations over Spanish entry, particularly the impact on the Community's self-sufficiency (and thus the level of agricultural expenditure) in certain agricultural products - Spain has a heavy citrus fruit surplus (250 per cent self-sufficiency) and is more than self-sufficient in most other fruit and vegetables. With the European Community as the biggest single fruit and vegetable importer in the world and Spain the largest single exporter, these high levels of self-sufficiency have already contributed to the rise in the Community's self-sufficiency in citrus fruits from 70 per cent in 1980 to 76 per cent in 1986/7 and from 145 per cent to 182 per cent in processed tomatoes over the same period.

Of all the potential benefits which accession offers to Spain, the financial transfers and expansion of agricultural trade through the adoption of the CAP offer the greatest attraction. Table 3.7 illustrates the importance of agriculture in the Spanish economy (though less than in Greece and Portugal), with some 15 per cent of the active population employed in agriculture and 5 per cent of GDP emanating from that sector. In relation to the Community, Spanish entry increased the total agricultural population by 17 per cent, the number of holdings by 22 per cent and total agricultural output by around 11 per cent (Commission 1988).

While large farm units predominate in the South, Spanish agriculture in general is typical of the Mediterranean region in its structural imbalance, with 55 per cent of holdings less than 5 hectares in size. Moreover, income disparities are particularly marked in certain agricultural areas - the 'poverty pockets' as described by Rodriguez (1983).

However, price incentives for the more efficient fruit and vegetable producers of the South have led to increased production in this sector, which ha been reflected by a further improvement in the agricultural trade balance. Prior to accession, despite tariff barriers, less than 60 per cent of Spanish agricultural exports went to European Community countries. With the removal of trade controls and the introduction of higher levels of support, the proportion of intra-EC exports has risen to almost two-thirds.

This 'inevitability' was a major factor in prolonging the progress of negotiations on Spanish accession, with several member states, particularly those threatened by Spanish competition in fruit, vegetables and wine (France, Italy and the Netherlands) warning of the 'damaging' consequences of opening the Community doors to the heart of the Mediterranean. The position of those countries 'opposed' to Spanish accession is aptly summed up by the concluding remarks of a report published by the Conseil National des Jeunes Agriculteurs (CNJA), the French young farmers organisation, which stated: 'Spanish accession is a mistake for all...the conditions that would make the European Community enlargement a benefit for Europe are not present...when a ship sinks, it is better not to take on board more passengers' (cit. Rodriguez 1983).

However, while Spain has been granted its boarding pass, it is important, as Punset (1985) stresses, to note the terms of entry. In particular, while it will have a say in future policy decisions within the Community, Spain (like all new entrants) is obliged to respect the conditions laid down by the Accession Agreement, which establishes the stages through which new member states must proceed in order to make adoption of common policies, particularly the CAP, practicable. In the case of Spain and Portugal, this means that tariff barriers will not be fully removed (or prices fully aligned) until 1996, thus easing the 'shock' to the Spanish economy, and the introduction of the Supplementary Trade Mechanism (STM) will enable both parties to the Accession Treaty to regulate the flow of trade in certain 'sensitive' products during the transition period (see Agra-Europe 1985).

CONCLUSION

The development of the CAP from the post war period to the early 1980s reflects the emergence of a wide range of contrasting and often conflicting pressures on the process of European integration. These pressures have reflected both the political constraints facing member states, in particular the ceding of sovereignty and the defence of national interests, and the economic consequences of adopting a common policy of open-ended price support in a sector which has responded beyond expectations.

Given the disparity between the agricultural sectors of the six, nine and ultimately twelve member states, it is perhaps surprising that the basic principles and mechanisms of the CAP have remained largely intact during its 30 years of evolution. The period from 1985 to the present day has seen a further growth in the pressure for CAP reform, both within the Community, as the cost of agricultural support has continued to rise, and from outside the Community, as the struggle to maintain exports markets in the face of subsidised production has also intensified. Yet the CAP remains as the single most important common policy, accounting for the lion's share of the Community budget and symbolic of the overwhelming desire to achieve common solutions to the Community's problems.

That there is a common policy for agriculture in the Community of twelve is as much the result of the importance which the CAP has attained as a symbol of European unity as it is of the desire of member states to respect the objectives for agricultural policy as outlined in the Treaty of Rome. In this respect the developments of recent years, namely the adoption of the Single European Act, the quest for economic and monetary union and the key role of agricultural policy reform in the Uruguay round of the GATT, may all serve to reduce the symbolic importance of the CAP and thereby facilitate, if not necessitate, effective rather than cosmetic changes to the way in which the Community chooses to support its farming population.

REFERENCES

Agra-Europe (1985) *Spain and Portugal in the EEC: the Mechanics of Accession*, Special Report No. 26, Agra-Europe, London.

Broad, R. and Jarrett, R. (1972) *Community Europe Today*, Oswald Wolff, London, pp 161-171.

Buckwell, A., Harvey, D., Thomson, K.J. and Parton, K. (1982) *The Costs of the Common Agricultural Policy*, Croom Helm, London, pp 25-66.

Clerc, F. (1979) Attitudes françaises vis-à-vis de la politique agricole commune. In: Tracy, M. and Hodac, I. (eds) *Prospects for Agriculture in the European Economic Community*, College of Europe, Brussels, pp 353-363.

Comité Intergouvernemental (1956) *Rapport des chefs de délégation aux ministres des affaires etrangères* (The Spaak Report), Brussels.

Commission of the European Communities (1958a) *First General Report on the Activities of the Community*, Brussels.

Commission of the European Communities (1958b) *Receuil des documents de la conférence agricole des états membres de la communauté économique européenne à Stresa au 12 Juillet 1958*, Brussels.

Commission of the European Communities (1968) *Memorandum on the Reform of Agriculture in the European Community*, COM (68) 1000, Brussels.

Commission of the European Communities (1973) *Improvements of the Common Agricultural Policy*, COM (73) 1850, Brussels.

Commission of the European Communities (1975) *Stocktaking of the Common Agricultural Policy*, COM (75) 100, Brussels.

Commission of the European Communities (1979) *Changes in the Common Agricultural Policy to Help Balance the Markets and Streamline Expenditure*, COM(79)710, Brussels.

Commission of the European Communities (1988) *Annual Situation in the Community: 1988 Report*, Brussels.

Cravinho, J. (1983) Portugal: Characteristics and Motives for Entry. In: Sampedro, J. and Payno, J. (eds) *The Second Enlargement of the European Community: a Case Study of Greece, Portugal and Spain*, Macmillan, London, pp 131-147.

De Abreu, A.T. (1983) Portugal: the Agricultural Sector. In: Sampedro, J. and Payno, J. (eds) *The Second Enlargement of the European Community: a Case Study of Greece, Portugal and Spain*, Macmillan, London, pp 149-165.

De la Mahotière, S. (1970) *Towards One Europe*, Pelican, London, pp 139-175.

De Vries, J. (1975) Benelux 1920-1970. In: Cipolla, C. (ed) (1978). *The Fontana Economic History of Europe*, Vol. 6, No. 1, Fontana, London, pp 55-63.

Efstratoglou-Todoulou, S. (1983) The Impact of the Common Agricultural Policy on the Greek agricultural Sector. In: Sampedro, J. and Payno, J. (eds) *The Second Enlargement of the European Community: a Case Study of Greece, Portugal and Spain,* Macmillan, London, pp 85-104.

European Communities (1987) *Treaties Establishing the European Communities, Treaties Amending these Treaties, Single European Act, Resolution-Declarations,* Office for Official Publications, Vol. 1, pp 207-257 (Articles 1-43 of the Treaty of Rome), Luxembourg.

Fearne, A. and Perry, S. (1988) UK Agriculture and the European Monetary System, *Farm Management,* Vol. 6, No. 12, pp 521-528.

Fennell, R. (1987) *The Common Agricultural Policy of the European Community,* BSP Professional Books, Oxford, pp 1-15.

Haagerup, N. and Thune, C. (1983) Denmark: the European Pragmatist. In: Hill, C. (ed.) *National Foreign Policies and European Political Co-operation,* Allen and Unwin, London, pp 106-118.

Harris, S. Swinbank, A. and Wilkinson, G. (1983) *The Food and Farm Policies of the European Community,* Wiley, Chichester.

Harvey, D.R. (1982) National Interests and the CAP, *Food Policy,* Vol. 7, No. 3, pp 174-190.

Heath, E. (1982) Interview in: *Europe 82,* No. 4, p 19.

Jordan, J. (1985) *The implications of Spain's accession to the EEC and the citrus fruit sector,* unpublished MA Dissertation, University of Reading, pp 3-10.

Kindleberger, C. (1965) The Post -War Resurgence of the French Economy. In: Hoffman, S. (ed) *In Search of France,* Harper and Row, London.

Kitzinger, U. (1967) *The European Common Market and Community,* Routledge and Kegan Paul, London, pp 33-198.

Koester, U. (1977) The Redistribution Effects of the Common Agricultural Financial System, *European Review of Agricultural Economics,* Vol. 4, No. 4, pp 321-345.

Lindberg, L. (1963) *The Political Dynamics of European Economic Integration,* Oxford University Press, Oxford, pp 261-283.

Mansholt, S. (1972) The Promised Land for a Community. In: Barber, J. and Reed, B. (eds) 1973 *European Community: Vision and Reality,* Croom Helm, London, pp 315-318 .

Marsh, J. (1979) United Kingdom Attitudes to the CAP. In: Tracy, M. and Hodac, I. (eds) *Prospects for Agriculture in the European Economic Community,* College of Europe, Brussels, pp 364-377.

Marsh, J. and Ritson C. (1971) *Agricultural Policy and the Common Market,* Chatham House, London

Marsh, J. and Swanney, P. (1980) *Agriculture and the European Community,* Allen and Unwin, London, pp 11-37.

Neville-Rolfe, E. (1984) *The Politics of Agriculture in the European Community*, European Centre for Political Studies, Brussels

Parker, G. (1979) *The Countries of Contemporary Europe*, Macmillan.

Payno, J. (1983a) The Second Enlargement from the Perspective of the New Members. In: Sampedro, J. and Payno, J. (eds) *The Second Enlargement of the European Community: a Case Study of Greece, Portugal and Spain*, Macmillan, London, pp 1-15.

Payno, J. (1983b) Spain - Characteristics and Motives for Entry. In: Sampedro, J. and Payno, J. (eds) *The Second Enlargement of the European Community: a Case Study of Greece, Portugal and Spain*, Macmillan, London, pp 187-207.

Pearce, J. (1983) The Common Agricultural Policy: the Accumulation of Special Interests. In: Wallace, H., Wallace, W. and Webb, C. (eds) *Policy-making in the European Community*, Wiley, London, pp 143-159.

Pepelasis, A. (1983) The Implications of Accession for the Greek Agricultural Sector. In: Sampedro, J. and Payno, J. (eds) *The Second Enlargement of the European Community: a Case Study of Greece, Portugal and Spain*, Macmillan, London, pp 70-84.

Punset, E. (1985) Limits of the Negotiations of the Adhesion of Spain to the EEC, *Anuario Hortofruiticola Espanol 1985*, pp 21-22.

Raup, P. (1970) Constraints and Potentials in Agriculture. In: Beck, R. *et al. The Changing Structure of Europe*, University of Minnesota Press, pp 126-170.

Ritson, C. (1979) An Economic Interpretation of National Attitudes to CAP Prices, Discussion Paper prepared for the second Wageningen seminar on: *The Role of the Economist in Policy Formulation with Respect to the European Community's Common Agricultural Policy.* University of Wageningen, Netherlands, December 13-14, 1979.

Ritson, C. (1983) The Economic Significance of the CAP *British Review of Economic Issues*, Vol. 4, No. 11, pp 1-15.

Ritson, C. and Tangermann, S. (1979) The Economics and Politics of Monetary Compensatory Amounts, *European Review of Agricultural Economics*, Vol. 6, No. 2, pp 119-130.

Ritson, C. and Fearne, A. (1984) Long Term Goals for the CAP, *European Review of Agricultural Economics*, Vol. 11, No. 2, pp 207-216.

Rodriguez, J. (1983) Spain: the Agricultural Sector. In Sampedro, J. and Payno, J. (eds) *The Second Enlargement of the European Community: a Case Study of Greece, Portugal and Spain*, Macmillan, London, pp 210-221.

Siotis, J. (1983) Greece: Characteristics and Motives for Entry. In: Sampedro, J. and Payno, J. *The Second Enlargement of the European Community: a Case Study of Greece, Portugal and Spain*, Macmillan, pp 57-67.

Swann, D. (1981) *The Economics of the Common Market*, Penguin, London

Tangermann, S. (1979a) Germany's Position on the CAP - is it all the Germans' fault? In: Tracy, M. and Hodac, I. (eds) *Prospects for Agriculture in The European Economic Community,* pp 395-404.

Tangermann, S. (1979b) Germany's role within the CAP: Domestic Problems in International Perspective, *Journal of Agricultural Economics,* Vol. 30, No. 3, pp 241-253.

Tangermann, S. (1980) National Attitudes to the CAP, unofficial translation by Rollo, J. of German article *'Agrarpolitische Positionen in den Mitgliedslanden der EG und den Institutionen'.* (unpublished).

Tracy, M. (1984) Issues of Agricultural Policy in a Historical Perspective, *Journal of Agricultural Economics,* Vol. 35, No. 3, pp 307-317.

Tracy, M. (1989) *Government and Agriculture in Western Europe - 1880-1988,* Harvester Wheatsheaf, Hemel Hempstead, pp 243-356.

Tsakaloyannis, P. (1980) The European Community and the Greek-Turkish dispute, *Journal of Common Market Studies,* Vol. 19, No. 1, pp 35-54.

Tsakaloyannis, P. (1983) Greece: Old Problems, New Prospects. In: Hill, C. (ed) *National Foreign Policies and European Political Co-operation,* Allen and Unwin, London, pp 121-134.

CHAPTER 4

BRITISH AGRICULTURAL POLICY UNDER THE CAP

Charles W. Capstick

INTRODUCTION

The task now facing British agriculture is how best to adapt to the stream of new policy initiatives issuing from Brussels. This process commenced in March 1984 with the announcement that milk quotas, requiring six per cent production cuts, would be introduced for the year commencing the following day. Since then there have been further quota cuts, beef support has been weakened, both wrung from the Community during the United Kingdom presidency, and in the spring of 1988 came stabilisers, set-aside, budget discipline and financial ceilings. After decades of false starts, commencing with Dr Mansholt's ambitious 1968 plan, it is abundantly clear that reform, this time, is for real. The changes so far signal for our agriculture a period of containment, possibly even one of overall contraction. And so it is said the curtain has been drawn, for a time at least, on some four decades when the goal of agricultural support policy was essentially expansionary. The strings were, of course, pulled in Brussels but it is right to add that it has for years been British policy to work for an end to the excesses of the Common Agricultural Policy.

This is a somewhat 'bearish' approach to the UK's experience of the CAP but it does, I believe, reflect at least a facet of the mood of the late 1980s. This is one of realism, of acceptance that the CAP was overdue for change and, following repeated postponements, better to start now than face worse later. But the other side of the coin is that the adjustment process provides both the opportunity to tackle the problems seen by many to be adjuncts to modern agricultural development and also, with 1992 on the horizon, to cash in on British agriculture's much vaunted competitive strengths. But what of the role played by our own policies during this transformation and can we discern any patterns for the future? Before turning to this task I will briefly examine the evolution of our policies during the CAP experience but avoid a round-by-round commentary. Rather, I shall pick out some key elements and inevitably bypass many important areas of policy such as trading arrangements, animal and plant health, food standards, marketing, research and the provision of extension services. I shall, however, introduce some thoughts on the way our own approach to agricultural policy has influenced the Common Agricultural Policy itself. As regards present issues, these I shall avoid, as it is not for an official to join in the public and political debates that now surround us.

PREPARING FOR ENTRY - THE FLYING START

The momentous step of applying the Common Agricultural Policy in this country was taken in February 1973. But the story commenced much earlier as it was declared policy to see that our agriculture should launch itself into Europe not from a 'standing' but from a 'flying' start. Well before the signing of the Treaty of Accession in January 1972, ratified later that year by Parliament, our joining was anticipated. Early in 1970 there was a distinct change in the tone of policy pronouncements and, following quickly, the level of support was also increased. It is worth recalling that up to then, during the 1960s, the principal economic preoccupations had been our relatively sluggish economic performance, recurrent balance of payments crises and the defence of sterling as if the whole ethos of the nation was at stake. Perhaps it was, but when sterling was devalued in 1967 prospective CAP prices came to appear even more extraordinarily generous. Agriculture during the 1960s was regarded as an important potential import saver and several policy statements stressed this by announcing versions of selective expansion plans. But these were invariably accompanied, in the fine print, by qualifications about expansion having to be based on productivity growth - which anyway was on a rising trend - and that much would depend on availability of resources and public funds for support. The problem was that import prices were well below support prices and as the Exchequer footed the bill for the difference the Paymaster was naturally wary. In the event, support price increases were modest and, although output tended to grow, this was largely from yield improvements.

Against this background economic assessments of the effects of adoption of the CAP published in February 1970 (UK Government 1970a) pointed to large support price increases and, above all, to an unattractive rise in food prices and a sizeable increase in the nation's food import bill. To add salt to the wound, a contribution to the agricultural side of the Community budget would bring the overall balance of payments cost to around £400 million [1] - a sizeable figure in those days. The message from this was clear. The bill arose largely from our low 'self-sufficiency' for food and if agriculture did not respond then it would ultimately have to be paid. Also, because CAP regimes meant that import prices would be CAP support prices (or world prices plus

1   It is interesting to note that at 1980 prices this figure would be equivalent to some £1,400 million, only a little less than the estimated actual cost to the UK of the CAP being borne at the turn of the decade.

levies, with the latter handed over to the Community budget), opportunity cost arguments against expansion lost their force.

The result of all this was a statement in early 1970 (UK Government 1970b) that 'Expansion is needed not only to save imports now but if we join the European Economic Community, to reduce the bill which would have to be paid through adoption of the CAP.' Support prices were indeed increased and further raised six months later. This was followed in 1971 (UK Government 1971), after entry negotiations had been proceeding for over a year, but with success not guaranteed, by an even more confident move. It was announced that 'The declared aim is to adapt the present system of agricultural support to one relying increasingly on import levy arrangements under which the farmer will get his return increasingly from the market.' This was largely, but not entirely, due to the imminence of the CAP as, by then, the deficiency payments bill was getting out of hand and heading for an unattractive 25 per cent rise the following year. 'The rising trend of production' - it was announced - 'was of public importance in connection with EC entry.' The policy green light was therefore shining even more brightly as incomes rose and with them investment. Asset values got the message as land prices doubled between 1971 and 1973 (see Figure 4.1).

THE TRANSITION YEARS

The 'best laid plans...' to smooth our adoption of the commodity regimes were documented in detail in the Treaty of Accession. There would be six price steps ending in 1977, whilst our contributions to the Community budget would not be paid in full until 1980. The reasoning behind all this was to spread the inflation effects, avoid awarding unnecessarily high producer prices to farmers, who would take time to respond with higher output, and finally to delay the financial burden for as long as practicable. It was hoped, not without good reason, that spending by the Community on agriculture would be controlled, so limiting the eventual financial burden. As the years went by such hopes were to evaporate and the famous words, penned in 1972 by the Community, that if unacceptable situations should arise 'The very survival of the Community would demand that the institutions find equitable solutions' eventually had to be brought into the equation.

Other events then broke over these detailed plans. The ending of the Bretton Woods system and the floating of sterling (outlined in the previous chapter) had meant that the 'green' pound would be accompanied by variable MCAs, correction factors and so on, that have come to test our mathematical aptitudes ever since. As sterling was weakening the green money system served, however, to return to British policy-makers some, albeit limited, autonomy over the level of CAP support prices in the United Kingdom. However, between 1972/73 and 1973/74, world cereal prices doubled, meaning that our producers experienced much higher prices than ever were envisaged, a matter of months after our adoption of the CAP. World sugar prices followed and soon after the world experienced the first round of world oil price increases. Inflation in the United Kingdom accelerated.

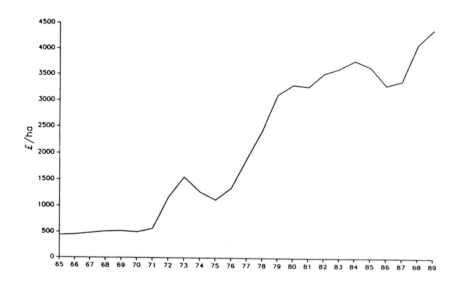

Figure 4.1: Agricultural Land Prices: England: All Sales

These events called for policy innovations on a major scale, generally to enable agriculture to progress in a balanced way. The arable sector prospered whilst the livestock side faced enormous feed cost increases in addition to those deriving from higher oil prices. The Government allowed fertiliser subsidies to be phased out but held on to calf subsidies until the end of the transition period to boost the beef sector. The pig sector was awarded a special subsidy in 1974/75 to help it through its ordeals but this was soon removed as being incompatible with the Community's rules of competition. Not to be outdone the Government tried again a year later when the sector was once again under pressure. Horticulture was also assisted with a subsidy on fuel oil.

Food prices were by now rising alarmingly; by 18 per cent in 1974 and 24 per cent in 1975 (see Table 4.1). Much of the blame inevitably fell, albeit quite

Table 4.1: Retail Food Price Index

	All Food Index	All Items Index
1965	14.5	14.8
1966	15.4	15.4
1967	15.4	15.8
1968	16.1	16.5
1969	17.1	17.4
1970	18.3	18.5
1971	20.3	20.3
1972	22.1	21.7
1973	25.4	23.7
1974	30.0	27.5
1975	37.6	34.2
1976	45.2	39.8
1977	53.8	46.1
1978	57.6	50.8
1979	64.5	56.7
1980	72.3	66.8
1981	78.4	74.8
1982	84.5	81.2
1983	87.2	85.0
1984	92.1	89.2
1985	95.0	94.6
1986	98.1	97.8
1987	101.6	101.9
1988	104.6	106.9
1989	110.5	115.2
1990	116.9	120.4

unfairly, on the Common Agricultural Policy. To help curb inflation - in effect to reduce the retail price index - the Government in 1974 introduced a new package of food subsidies which included bread, flour, milk, butter, cheese and tea. Although phased out by 1977 their cost rose to over £600 million in 1975. The Community acquiesced as the subsidies did serve to increase demand, particularly for milk products, which by then were in surplus. To add to all these economic uncertainties, the new Labour Government, less enthusiastic about membership than its predecessor, was committed to renegotiate the terms of entry and also to conduct a referendum on our continued membership. The background to all this is a subject for the political historian but the former brought little material change on the agricultural front, save for the partial reintroduction of deficiency payments in the form of a beef premium, whilst the referendum in 1975 strongly endorsed our membership. But these events did indeed create an air of uncertainty for our farming sector for some time.

Faced with rapid inflation, a depreciating currency, large negative MCAs, modest but steady output growth and a varying pattern of subsidies to provide relief to both farmers and consumers, the Government decided it was time to present its view on the prospects and set out its policy for agriculture looking to the end of the decade.

Food From Our Own Resources (UK Government 1975) was amongst the most analysed and discussed - not all of it complimentary - of such documents for many a year. Ignored it was not. Essentially its intention was to show that in spite of the economic turmoil of the period there was, through it all, an enduring vision of an expanding industry. The document was almost exclusively concerned with the production of commodities. This was understandable, since the cost of food imports had doubled in two years (see Figure 4.2) and the prospect was for imported food prices to remain firm. Noting the industry's record of increasing efficiency, the Government declared that it could see the production of the industry growing at around 2.5 per cent per annum to the end of the decade. Most of the main commodities would contribute but cereals, milk, beef and sheepmeat would play the major roles. Unusually, projections were published and attention inevitably focused on their realism. Agricultural policies would be framed in the light of the foreseen expansion and later in the year the 'green' pound was devalued sufficiently to raise agricultural prices by about 11 per cent and milk prices were also increased. But barely had the ink dried before farmers were faced with the droughts of 1975 and 1976, which severely reduced production. It was not to return to its trend growth until 1978.

One point worth noting here is the emphasis, or rather lack of it, placed on environmental issues. The White Paper did, however, comment on concerns about pollution and the changing appearance of the countryside but essentially to assert that the production expansion 'should not result in any undesirable changes in the environment'. Although such a statement was hardly likely to attract applause from environmentalists, the mere mention of the problem was of some significance whilst its virtual dismissal should be seen very much in the context of the economic battles being fought during those years. I shall return to this topic below.

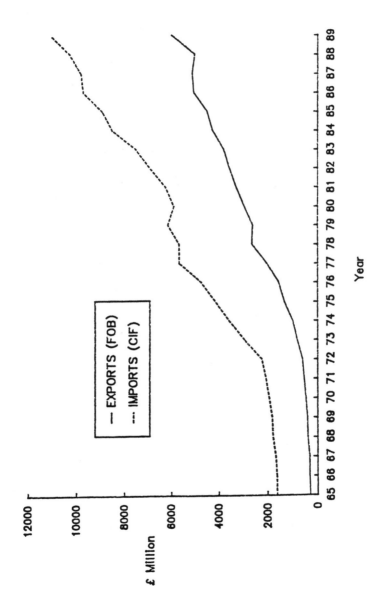

Figure 4.2: Value (£ million) of Total Trade in Food, Feed and Beverages

Mention should be made also of several initiatives taken by successive Governments to preserve features of our old support system held in particular esteem. One such example is the scheme to provide special aid to agriculture in our hills and uplands. The ability to continue this was secured during the entry negotiations and eventually our arrangements were incorporated into the Community's own 1975 Less Favoured Areas scheme. Today farm incomes in areas suffering various disadvantages continue to be enhanced and the principle has readily been taken up by most other member states. Indeed, something like a third of the European Community agricultural area now comes under the Less Favoured Areas umbrella whilst in this country the proportion is about a half.

On the milk front, concerns emerged during the transition period about the place of the Milk Marketing Boards in the Community scheme of things. Their size, the complex price-setting machinery for liquid and manufacturing milk, and their role as sole buyers of milk, had generated much unease in Community quarters. But new arrangements were introduced, the Community was satisfied, and following a referendum of producers, who voted overwhelmingly to retain the Boards, their future was secured. On another flank the United Kingdom support system had for decades included aids for investment and the modernisation of farms in the form of capital grants. These were also preserved in their essential form, and whilst under the Community scheme most other member states paid the aid in the form of interest rate subsidies, it remained United Kingdom policy to retain grants. There were several sound reasons for this, not least our arrangements seemed to work well and anyway there was no tradition in this country of paying interest rate subsidies. Grants meant that farmers had no need to borrow in order to receive a specified amount of aid. Much investment in the 1970s derived from internally generated rather than from borrowed funds, a pattern that was soon to change.

Several schemes such as fertiliser, lime and calf subsidies, together with guaranteed price arrangements, were all eventually phased out, the benefits they provided being taken over by the generous CAP commodity regimes. In this way successive Governments managed to retain some essential features of our pre-accession agriculture whilst at the same time running the CAP regimes (but phasing in the price levels) much as they had been inherited.

PRICES, GREEN MONEY AND BUDGET COSTS

The Community experienced its first taste of surpluses in the late 1960s. There had then been a lull but by the time of our entry the search was on for export outlets, particularly for butter. Later, several sets of reform proposals to cope with expanding production emerged from Brussels. Some elements of these were adopted and accepted by the British Government, such as subsidies on school milk and butter - enabling continuation of our own scheme - and various retirement and dairy herd conversion schemes. But the main cause and the key

British preoccupation was that Common Agricultural Policy prices were too high. Indeed real support prices in the Community were not declining and, given productivity growth, the incentive was there to expand. It did so. The reform proposals were therefore invariably pushing against market signals.

For the British, not only had CAP prices always seemed high, but they had been increased during the transition period and we were destined to move towards them. Also the green pound was more highly valued than the sterling market rate and this provided for us further potential price increases. The Government's policy, however, was to devalue the green pound only when this was judged necessary and even then only by modest portions. This was largely designed to avoid excessive price increases for farmers when other sectors were facing quite severe price and income restraint policies and also to avoid the inflationary effect of higher food prices. Agriculture could not be made an exception and consequently a large gap developed between the green and market rates. The industry, however, was to reap the benefit from this subsequently.

On the budgetary front the costs of the CAP were beginning to rise inexorably. Although not a direct element of policy towards agriculture, the increase in the agriculture side of the Community budget from £1,900 million in 1973 to £6,000 million in 1978 and the inevitability of massive burdens for the United Kingdom by 1980, after the end of the financial transition period, were all bound to have repercussions (see Figure 4.3, which shows the growth of Community spending on agriculture expressed in European Currency Units). The Government accordingly set out to invoke the 'unacceptable situations' clause and, after some interim arrangements, these efforts finally culminated in the 1983 Fontainebleau rebate system. This served to reduce the otherwise large UK subsidy to the rest of the EC but it had further important consequence. Prior to it, only two member states made large net contributions, namely Germany and the United Kingdom. Others therefore tended to benefit financially from higher CAP spending and were unlikely to endorse CAP price cuts: some even pressed strongly for price increases. The Fontainebleau system succeeded in spreading the financial burden more evenly and this has certainly engendered a more responsible approach to price determinations.

THE POST-TRANSITION PHASE

This was the period when the volume of UK gross agricultural output truly expanded, rising by over 20 per cent between 1977 and 1984. At the Community level production, stockpiles and spending on agriculture annually reached new heights such that eventually agricultural issues frequently passed from the Agricultural Council to occupy much of the agenda of meetings of Community Prime Ministers. But it was also the period when our own deep concerns, on the one hand, with the excesses of the CAP, became, on the other, inextricably interlinked with concerns for the environmental impact of modern and expanding agriculture. The relationship between the two, with the need to

Figure 4.3: FEOGA Guarantee Expenditure 1973-1990

curb production at the Community level, yet ensure that agriculture provides environmental goods, and remains competitive, is complex and still evolves. But I am in no doubt that present policies, which are striving to reconcile the two, have their origins in the immediate post-transition period.

This is perhaps best illustrated by reference to the 1979 White Paper *Farming and the Nation* (UK Government 1979a). This differed markedly in tone from the earlier *Food From Our Own Resources* although it continued to emphasise that 'a substantial increase in agricultural net product is in the national interest and can be achieved'. No targets or commodity priorities were set but it foresaw a 10-20 per cent increase in output by 1982/83. In the event this was safely achieved. Also outlined was the Government's approach to the problems of the CAP, with the commitment reiterated to seek to 'restrain' common price levels. Although a cautious phrase, this did, of course, mean cuts as is clear from the rest of the document. For our producers there was always the green pound card to put in play and strong hints were given of future devaluations each being '...carefully judged on its merits at the time'.

Significant sections of the White Paper were devoted to the question of the balance between agricultural productivity and environmental concerns. It spoke welcomingly of various studies searching for ways of retaining certain desirable aspects of the countryside and recognised that in some cases agriculture would have to take second place. This would be most apparent in accepting applications for capital grants where amenity aspects would have to be taken into account. In the case of drainage works, for example, account would have to be taken, as required by the Water Act 1973, of the impact on the environment and conservation. Soon after the White Paper's publication the agricultural report of the Royal Commission on Environmental Pollution (UK Government 1979) emerged to bring such matters further into focus. It is therefore to disregard history to assert that the 'Greening of the MAFF' is only a recent phenomenon. There was for many years a distinct greenish hue but it must also be recognised by observers today, when times are more prosperous, that during periods of economic turmoil agriculture was called upon to produce, and produce ever more efficiently, and it was for the Ministry of Agriculture, Fisheries and Food to devise the policies to achieve those ends. Inevitably conservation issues had to be relegated in the list of priorities.

Farming and the Nation it must be said did not attract lasting attention, largely because in a matter of a couple of months there was a change of Government. The green pound was then soon devalued to give agriculture a boost and, whilst eroded by inflation, production was expanding. At this time the Community decided to bring mutton and lamb into the family of commodity regimes and with them the word 'sheepmeat' entered our vocabulary. It was the one product for which our production would be dominant in the Community, so opening up the prospect at long last of something for which our budget receipts could be greater than our contributions. The chance was not missed and the regime negotiated for this

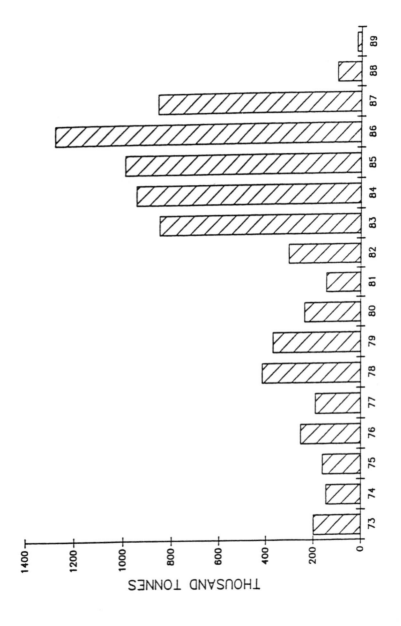

Figure 4.4: EC Butter Intervention Stocks

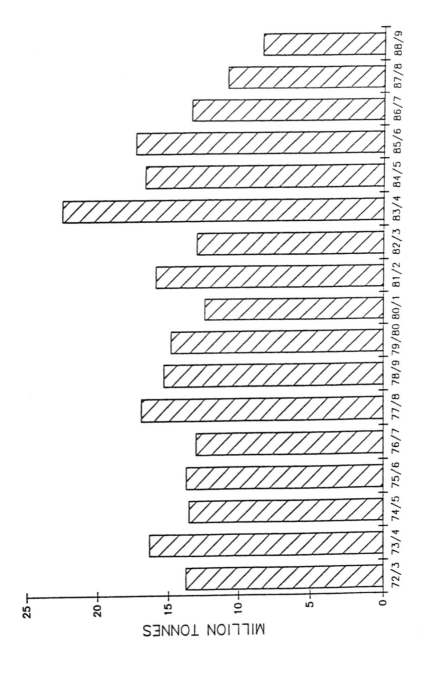

Figure 4.5: EC Cereal Intervention Stocks

country was based on variable deficiency payments or premiums, although other countries chose to avoid variable premiums and were protected by 'clawback' arrangements from the lower market prices here. Considerable financial and producer benefits derived from this accord since when our sheep flock has expanded, the growth now being based very much on exports. More recently, the Community has started to phase in new arrangements with a single ewe premium.

It was around the turn of the decade that several long-established parameters changed markedly. At the Community level, production was expanding, and so were stockpiles and budget costs - a triumvirate carved in the souls of all CAP aficionados. In this country the strengthening of sterling, coupled with earlier devaluations, turned negative into positive MCAs. Should the green pound therefore be revalued, so reducing producer prices? The Government's answer was 'no', as sterling was floating and who could tell where it would next navigate. This of course proved to be absolutely right as events later demonstrated.

In spite of the production projections, the Community increased support prices by 4, 7 and 10 per cent, in the years 1980, 1981 and 1982. Green rate devaluations around the European Community meant that in national currencies the increases were even greater. [2] The United Kingdom vetoed the 1982 increases but was ignored and they were put into effect. The consequences, namely rising production of milk and other products, soon came home to roost. Projections pointed to unsustainable stock and budget positions on the milk front and it was against this background that the milk quota card simply had to be played (see Figures 4.4 and 4.5 showing intervention stock levels for butter and cereals over a 17-year period). The time for using price policy was by then long past. As I indicated at the outset, there have since been other measures for a wide range of products together with the introduction of much greater budget discipline. In spite of generally welcome developments, the costs of the CAP, both economic and budgetary, are presently and seem destined to remain substantial. Yet it is true that intervention stocks have recently fallen markedly. Therefore much scope remains for further reform measures and it is British Government policy to press that progress be maintained.

NEW DIRECTIONS FOR POLICIES

I have concentrated so far on mainstream, essentially commodity and price, policies but there is of course much more. In these concluding paragraphs I shall touch on these wider issues where the approach being developed in this country has been influential on the Community front. Mention has already been made of the way environmental issues in the 1970s tended to be

2   This issue is discussed in detail in Chapter 6.

overwhelmed by economic preoccupations and how the balance was changing such that today they have become key elements of policy. This is best illustrated by the Government's 1987 series of publications *Farming and Rural Enterprise* (UK Government 1987). Prior to this, academics and others had issued a stream of pamphlets and polemics drawing attention to the concurrence of their worries about the countryside, pollution and the rural economy with production in excess of market demand. Why not, they said, use the funds now disbursed on storage and export subsidies to encourage farmers to produce 'environmental goods' that are demanded by the taxpayers and consumers?

Such arguments were in fact pushing at a door that was already opening. The demand for these non-food goods, for which the term 'CARE goods' has been coined by Professor McInerney (1984) - representing conservation, amenity and rural environment - tends not to be satisfied by normal market mechanisms. Consequently, if they are to be supplied in greater abundance then, so the argument goes, Government intervention will be necessary. There is nothing original here as our policies towards farming in the hills and uplands have been operating for many years, designed as they are to encourage the retention of the economic and social structure in these remoter areas. More examples have now been added with the creation of environmentally sensitive areas (ESAs), endorsed by the Community in 1985, which enable farmers to be paid for retaining and therefore supplying particular farming systems considered desirable. Nineteen such areas have been declared, encompassing three-quarters of a million hectares.

If we accept that continued expansion of Community production cannot be sustained but that productivity will continue to advance, it follows that some land will be less intensively farmed. The challenge is to provide the conditions for economic activity to be maintained and possibly expand in such a climate. Diversification grants are now available, the experimental Farm Woodlands Scheme is attracting takers whilst other Government agencies provide assistance, for example, to protect areas of special interest and for the creation of new rural enterprises, whether based on or off farms. Whilst the new set-aside scheme for arable land is first and foremost an adjunct to price policy towards cereal and other crops, under it income aid will be provided for those who find they have hectares that have become unprofitable. The principles of this scheme were first proposed by the British Government in 1986 and now it is operating across the Community. Conditions are attached which require that set-aside land should be maintained in an environmentally satisfactory manner but it is, essentially, a scheme to assist farmers to adapt to the economic changes now facing the industry. On the conservation front, the Wildlife and Countryside Act provides for management agreements, with compensation, to allow landscape and habitats to be preserved. Additionally, the Ministry of Agriculture makes available grants such as for hedgerows, footpaths and so on, to improve the rural landscape and access.

I quote these few examples not to imply that the problems facing agriculture as it adjusts to curbs on expansion will be comprehensively overcome by a multitude of schemes. What they do confirm is that agricultural policy embraces far more than prices and commodity regimes, crucial though these remain. The emphasis has transparently changed with practical recognition of the need to satisfy public demand for CARE goods and at the same time reward farmers for supplying them. But the scale and nature of that demand is not amenable to easy quantification and no easy formula exists guaranteeing that the right balance can quickly be struck between incentives and supply. Whilst a key principle of policy is to encourage farmers to participate on a voluntary basis, this does mean that new policy initiatives have to be tentative and probing as well as innovatory. They will be evaluated and modified in the light of experience gained.

CONCLUSION

Between 1973 and 1978 the volume of United Kingdom agricultural gross output increased by over a fifth, using 1.25 million fewer hectares of crops and grass and with 100,000 fewer workers. Our self-sufficiency for indigenous-type foods has risen from 60 per cent to over 80 per cent in 1984 but due to milk quotas and poorer crops this subsequently fell to around 75 per cent. Although largely deriving from technical advances this progress was powerfully underwritten by CAP support policies as they applied in this country. But the other side of the coin tells us that it has been an expensive experience in terms of the financial and economic resources diverted to agriculture and even now the cost remains high. The reform process, now commenced and so long advocated by the United Kingdom, remains to be completed.

One could look back and wish that certain decisions had been different, that the price increases in the early 1980s had been brushed aside, and that a neutral, rather than the strong, currency system for green money had been agreed in 1984, and so on. But perhaps at those times the Community, excepting, dare one say, the United Kingdom, was neither economically nor politically ready to launch itself into the reform process. It has now commenced the task but is surely facing greater adjustments than would have existed a few years ago. Also, in addition to domestic pressures for change are to be added those coming from the outside world where GATT is the forum in which demands for world-wide reduction in support - multilateral subsidy disarmament - are negotiated.

The inevitability that the Common Agricultural Policy would have to adapt has long been evident but I would suggest that for some years the search has been on in this country for ways in which our own policies might be changed. If production of traditional products were unable to continue to expand whilst yields grew, then some land would be less intensively farmed. If the revenue of the industry could not grow as in the past, as support prices are held or reduced, then that revenue would be unable to support the same number of farmers and workers to acceptable standards.

These are straightforward equations and, by acknowledging them, British policy can already be seen to be placing greater emphasis on the encouragement of new enterprises both agricultural and non-agricultural and the stimulation of the industry to produce environmental or 'CARE' goods in even greater abundance. The process will surely continue to evolve. But it has also been a basic tenet of policy that the industry should be able to compete on equal terms with the rest of the Community. The 'level playing field' has been the aim whether on labelling, health standards, marketing or support. The prospect of the unified market after 1992 and phase-out of MCAs brings these issues into even greater focus and is now a key preoccupation. Beyond this time scale I would be travelling deep into the area of clairvoyance. But I shall not do so as our dear friend John Ashton would surely have disapproved of economists trespassing too far into that kind of territory.

REFERENCES

McInerney, J. (1984) *'Agricultural Policy'*, Inaugural Lecture, University of Exeter.

UK Government (1970a) *Britain and the European Communities, An Economic Assessment*, Cmnd 4289, HMSO, London.

UK Government (1970b) *Annual Review and Determination of Guarantees 1970*, Cmnd 4321, HMSO, London.

UK Government (1971) *Annual Review and Determination of Guarantees 1971*, Cmnd 4623, HMSO, London.

UK Government (1975) *Food From Our Own Resources*, Cmnd 6020, HMSO, London.

UK Government (1979a) *Farming and the Nation*, Cmnd 7458, HMSO, London.

UK Government (1979b) Royal Commission on Environmental Pollution, 7th Report: *Agriculture and Pollution*, Cmnd 7644, HMSO, London.

UK Government (1987) *Farming and Rural Enterprise.* HMSO - incorporating *'Farming UK'*, *'Rural Enterprise and Development'* and *'Rural Scotland'*.

CHAPTER 5

MODELLING THE CAP

Kenneth J. Thomson

INTRODUCTION

John Ashton was many things, but not perhaps a modeller. He took a humanistic and political approach to economics, and neither the elegance of mathematical language nor the numerical dexterity of computers held much appeal for him. Thus econometric modelling, whether of the earlier linear programming and regression types, or the more recent simulation and general equilibrium varieties, was regarded by John, one suspects, as activities for backroom boffins to busy themselves with, rather than the proper and public face of agricultural economics.

But the Common Agricultural Policy was a different matter to John, who had been involved in Britain's first application to join the European Community in 1963. He saw, with perhaps a mixture of excitement and foreboding, that this country's accession to the European Community in 1973 would bring about a fundamental structural change to the farming environment of the United Kingdom. To this end, he actively sought and encouraged younger, more technocratic colleagues who would attempt to apply quantitative techniques to this massively important problem. Perhaps he found it no easier than the rest of us to manage such research, in the sense of assessing the resources required for desired computational achievements, or judging the probabilities as to which, if any, of those achievements would yield useful and interesting findings. But that is a task that can be faced with equanimity only by ministers of education and chairmen of research councils.

This paper attempts an overview of the development - dare we call it progress - of quantitative modelling of the CAP over the last two or three decades. It should be made clear that this is not the same thing as the development of economic analysis of the Common Agricultural Policy. The fertile minds of politicians and their advisers have come up with a succession of *ad hoc* measures to deal with the not infrequent crises of European agriculture: the long list includes MCAs, co-responsibility levies, quotas, hormone bans and now set-aside. Economists, though occasionally hard-pressed, have come up in response with relevant and numerical analyses of the measures and their probable consequences, but usually without the assistance or otherwise of the fully econometric models dealt with later.

If modelling is not the whole of economic analysis, it is even less the analysis of the development of policy itself, or of the views taken of policy by governments and others over the years. Here, economic contributions can certainly be made, both by public-choice theorists (for example Hagedorn, 1983, Wallace 1989) and by agricultural economists themselves (for example Heidhues 1976, Langworthy *et al.* 1981, and recently Fearne 1989). But the political agenda is not yet set by laptop computers, and the need for clear economic thinking and expression is more important than yet more sophisticated calculations. Perhaps the time will come when modelling insights - both conceptual and numerical - will serve to assist rather than confuse our policy-makers - but not yet!

THE NATURE OF MODELLING

By modelling is meant here the representation of a system - all or part of the European Community agricultural industry in our case - by a set of equations applied to data in order to calculate the outcome of a number of situations (Table 5.1). The equations may be directly estimated from observed data in the classical econometric approach. Alternatively, their parameters may be chosen more or less indirectly by the modeller in the light of past estimation efforts or his/her own judgement. Some people are of course acutely unhappy when coefficients are thus assigned their values, pointing out that all kinds of theoretical conditions may be broken. However, when pressed, the objectors often have to confess that their logic depends on a number of rather improbable conditions - such as constant returns to scale, or the maintenance of market equilibrium - which make that viewpoint less than easy to defend when non-academics are listening, and in any case all models will be judged by non-economists by their plausibility in performance rather than their theoretical basis.

Table 5.1: Definitions of Modelling

I. The representation of a system by a set of equations applied to data in order to calculate the outcome of a number of different situations

II. What modellers do:	mathematics
	computing
	economics
	agricultural science
	communications
	management

The general development of computing facilities, and of economic theory expressed in algebraic terms, has underlain the general fashion for modelling in economics, as in other disciplines, over the last few decades. However, there are a number of special features which may distinguish agricultural policy modelling.

First, of course, like agricultural policy itself, there is a lot of it about. That is partly because - for good or bad reasons - there are a lot of agricultural economists about, relative to the size of our industry. Then, there are available a lot of data on farming: not such a good thing from the point of view of the modeller as one might think, but certainly an incentive towards quantification. And because the governments of the industrialised world have seen fit to operate agricultural policy largely by intervention in the market-place - by deficiency payments, tariffs, and subsidies on exports and inputs - rather than by other means such as regulations over input usage, or targeted investments in farming's human capital, econometric models have been fairly obviously applicable. The CAP has been conspicuously open to analysis in this respect, as shown by the well-known statistic that 95 per cent of the Community's agricultural budget is spent on the so-called Guarantee Section devoted to price support, as opposed to the remaining 5 per cent devoted to the Guidance Section for structural expenditure. This point is returned to near the end of the chapter.

Another interpretation of modelling is the day-to-day activities of the modellers themselves (see McClements 1973, Thomson and Rayner 1984). This is not a well-explored field, but the appropriate combination of expertise in mathematics, computing, economics, agriculture, communications and plain old management seems to be of crucial importance in successful modelling efforts. One obvious aspect is persistence: a model constructed during the last few months of a PhD is unlikely to enjoy a long and useful maturity as its progenitor moves on to pastures new, and the more successful macroeconomic models seem to be based on a solid and secure institutional foundation, with a long-term role for the model itself.

CLASSIFYING CAP MODELLING

In a paper some years ago (Thomson 1982), I ventured to classify model use in policy analysis into three categories:

a) **'strategic'**, addressing changes *of* policy, with the emphasis on broad investigation along relatively novel lines for wide and often lay consumption, and often based on a rather small data base;

b) **'tactical'**, evaluating changes *in* policy, within a non-controversial analytic framework with quite often considerable quantities of data, in order to compare alternative settings of policy instruments, for professional and industrial consumption; and

c) **'operational',** to assist a decision-making policy process, such as stocks control or export subsidisation, executed by governmental agents, and of course using actual and up-to-date data.

Naturally, the boundaries between these are not well-defined or watertight, and given that the CAP itself is highly controversial, let alone the appropriateness of its support price levels or the efficiency of the Commission's activities, it is not surprising to find models encompassing a number of aspects from each of these categories, especially the first two. Nevertheless, it may be worth bearing in mind the classification system (which is summarised in Table 5.2) as we examine a few examples of CAP modelling.

Table 5.2:   Model Use in Policy Analysis

Type	Strategic	Tactical	Operational
Application	Change of Policy	Changes in Policy	Execution of Policy
Approach	Creative	Evaluative	Specific
Audience	Public	Expert	Administrative
Data Base	Small	Large	Available
Estimation	Rough	Intermediate	Sophisticated

SOME EXAMPLES OF CAP MODELLING

There have been a number of useful reviews of recent model use to analyse various aspects of the CAP, including the latter two Siena Workshops (Tarditi *et al.* 1989), and a recent EAAE Seminar on Agricultural Sector Modelling. At the latter, Bauer (1988) discussed agricultural sector models in terms of a number of characteristics (Table 5.3).

There are a number of historical roots from which current modelling work has grown. The work of Nerlove (1958) on agricultural supply has generated a host of more or less *ad hoc* regression-type models addressed to the fluctuations in production of various farm commodities. From the point of view of modelling the CAP, this approach has a number of drawbacks. First, the policy has been notably successful in maintaining market stability and advancing technical productivity. Both these tend to mask the adaptation process which the Nerlovian formulation is set up to simulate. Second, surpluses and budget expenditure, rather than supply in itself, are most usually the focus of CAP analysis, and this requires attention to the demand as well as the supply side in many markets. Third, there is the problem in the EC of covering a large number of commodities and several very different countries, some of which have adapted to significantly changed policy environments within a relatively short time-span.

Table 5.3:   Characteristics of Agricultural Sector Models

Coverage	single or multi-commodity
	sectoral or general equilibrium
Linkages	commodities
	subsectors
Complexity	linkages
	policy instruments
(Dis)aggregation	commodities (products, factors,
and Differentiation	and so on)
	countries or regions
	enterprise or farm groups
Equilibrium	product and resource markets
	static or dynamic ('adjustment')
Explanatory or	consistency
Forecasting Power	plausibility

Another component of our econometric heritage is the use of optimisation programming techniques associated with Heady and his followers. This approach promises to allow the incorporation of regional and transport features into a sectoral model, as well as the specification and analysis of farm resource use. It has been used quite extensively in analysing agricultural development plans for Third World countries (Hazell and Scandizzo 1986). However, for the purpose of modelling the heterogeneity of European agriculture, there seems to have been limited progress, as in a previous British context (Thomson and Buckwell 1979), or indeed, so far as one can judge, back in the United States itself.

In practice, therefore, more *ad hoc* approaches to CAP modelling have been most common, using a variety of equation types in an attempt to capture the main features of the major farming sectors or commodities. Some, such as the EC component of the International Institute for Applied Systems Analysis (IIASA) Basic Linked System (BLS) model (Fischer *et al.* 1988), are large and detailed. In others, though often based on detailed econometric work, more or less roughly calibrated coefficients have been used to construct a market-policy model covering most of the main commodities and the more important policy instruments (Buckwell *et al.* 1982, Mahé and Moreddu 1989).

A recent experience was the so-called 'Disharmonies' study convened from a group of European and American agricultural economists by the European Commission (1988a). This exercise attempted to bring to bear a number of existing models on a set of four pre-defined policy options for both the EC and the United States in advance of the GATT negotiations. There is no space to discuss this study here; the massive tome from Brussels can be pored over by those interested. But it was a salutary experience for the professionals engaged in trying to reconcile different data bases and model outcomes for an immediate and important customer. The EC Commission is now more active in seeking to find out whether policy models can assist it in its work; it is to be hoped that the Disharmonies study was a step forward rather than backwards in its evaluation of model usefulness.

CURRENT DEVELOPMENTS

Recent activity in modelling the CAP is being influenced by a number of factors. It is not clear in what order to place these influences: perhaps some will prove more enduring than others, while their importance can of course be measured in various ways. Also, one cannot be exhaustive, but four main influences are:

> International Trade Problems
> Duality Theory
> The Environment
> Socio-Structural Measures.

First, it should not be surprising in the latter half of the 1980s that international trade issues have been more important than for many years. Several well-known modelling exercises have been carried out specifically for negotiating parties in the current GATT round and, given the importance of the Community in world trade in (and especially trade disputes over) agricultural products, the EC and its CAP have been significant components in these constructions (Tyers and Anderson 1988, OECD 1987, Centre for International Economics 1988). Undoubtedly this modelling work has been influential in forming and changing attitudes towards the external aspects of the CAP. Whether this will be enough to force reform faster than is occurring under budgetary and other pressures remains to be seen.

Second, there have been a number of theoretical developments with modelling implications. One, though not to my knowledge applied as yet to the CAP as a whole, is duality theory, using cost and profit functions to handle a number of both inputs and outputs (see Glass and McKillop (1989)for a recent United Kingdom example). Another is general equilibrium analysis of the whole economy, derived from the original Leontief input-output approach. This has generated a number of formulations, including the BLS model already referred to, and a number of specifically European models (Keyser 1986, Munk 1984). The attraction here is consistency of model data and results, due to the comprehensive theoretical framework within which these computable general equilibrium (CGE) models are built. Again, its applicability to the more immediate problems of the CAP, with its complex policy and budget structure, has yet to be tested.

Third, the CAP is itself changing, though slowly, under the pressures of several influences, amongst which we must still place budgetary constraints first, but with environmental concerns catching up fast. As scientists generally, and economic modellers in particular, manage to obtain less patchy and unreliable measurements of the EC's rural environment, there is no doubt that attempts will be made to link these with the Community's farming industry and its policy measures. Indeed, such efforts have already been made, and with Newcastle academics involved. At first, no doubt, these CAP-environmental models will be crude adaptations of versions built primarily for market analysis, but with enough effort more specialisation may be expected.

Finally (in this catalogue), and related to the previous point, more serious modelling attention may be proper to the structural aspects of the CAP, concerning the size and pattern of farms, and their individual use of resources. Although certain regional farming sectors have for long been heavily dependent on Guidance Section spending, structural change in European agriculture has appeared somewhat resistant or indifferent to this policy approach, on which in any case expenditure has been minimal, for a number of reasons. But concerns over the social responsibilities of the European Community in rural areas, as expressed recently by the EC Commission (1988b), are likely to be translated into more active measures directing spending at farming. These measures accord more with the attitudes of urban and industrial man than the productively efficient agribusiness which past policies have undoubtedly promoted. In this case, modelling the CAP becomes an occupation somewhat different, more difficult, and perhaps less disinterested. If so, we shall miss even more John Ashton's gift for seeing beyond the methodology.

APPENDIX: THE NEWCASTLE CAP MODEL

The Newcastle CAP model began life in the late 1970s in an effort to provide some quantitative economic information on the then-fierce public debate on the financial cost to the UK of joining the European Community. By far the largest component of this cost was (and is) due to the budgetary burden of the CAP, coupled to the EC's revenue formulae linking national contributions to levies on farm products and to GNP. The small size of the UK agricultural sector relative to the British economy as a whole (itself larger relative to the Community GNP in the late 1970s than ten years later) meant that the UK found itself contributing a very substantial amount to the EC budget at that time, despite earlier efforts to renegotiate a more satisfactory agreement. This was, indeed, one of the main planks of the incoming Prime Minister, determined to recover 'Mrs Thatcher's billion'.

Although several estimates of the cost of the UK accession had been made prior to entry in 1973 (for example, Warley 1967, Butterwick and Neville-Rolfe 1968, HM Treasury 1970), these were seriously outdated by the late

1970s. Moreover, the 1970s were a decade in which unprecedented post-war instability in global agricultural, energy and financial sectors made efforts to track and predict economic developments extremely difficult. There were of course attempts to model the impact of the CAP on British farming, for example the so-called microeconomic supply model of British agriculture (Buckwell and Thomson 1979). But the more detailed budgetary aspects of the CAP were virtually ignored in the mid-1970s, when UK agricultural economists were in any case occupied with a combination of severe drought, high inflation and the 'green pound' issue.

Three UK publications provided an important spur to the Newcastle CAP project. One (Rollo and Warwick 1979) reported some relatively sophisticated calculations carried out at the Ministry of Agriculture in order to assess (presumably for Treasury purposes) the 'resource cost' of the CAP. Another (McFarquhar *et al.* 1977) was a Cambridge estimate of the 'food cost' of the CAP to the UK. The third (Morris 1980), from the Institute of Fiscal Studies, used economic calculations similar to those of the Newcastle model to arrive at resource costs of the CAP for each member state.

In comparison with these exercises, the Newcastle project, financed by research grants from the Social Science Research Council (later the Economic and Social Research Council) and the Ministry of Agriculture, Fisheries and Food, took a wider, all-EC approach, and tried to take greater account of the commodity and policy detail which by the late 1970s was becoming dominant as the complexities and financial burden of the situation became clearer.

The model essentially comprises a set of simple calculations corresponding to the comparative-static computation of market equilibria and welfare changes for each major CAP commodity in each EC member state. At the start, nine countries were included (Greece only joining the Community in 1980, and data being rather sparse), and the commodity list included wine, fruit and vegetables but not sheepmeat (for which a regime was introduced that year). The alternative policies with which the base position was compared included price-fixing changes in support prices, elimination of green exchange rates, exact EC self-sufficiency in each product, and complete elimination of the CAP. The model was extended in the early 1980s to include Greece (and, later, Spain and Portugal). The commodity coverage was altered to include sheepmeat and olive oil, and to exclude wine, fruit and vegetables on grounds of product heterogeneity. Commodity and policy details were also modified in line with real-world developments, notably the introduction of milk quotas in 1984. The 'rest-of-world' component, though never well developed, was refined to allow for price reactions on world markets to changes in the CAP, and calculations were added to reflect effects in the feedingstuffs markets and to analyse the distributional features of CAP alterations. An important extension was the compilation of a time-series data base from 1976 onwards, and the programming of an extrapolation subsystem to provide current or future 'data' on which to run the model itself.

The initial stage of the project ended with the production of a book (Buckwell *et al.* 1982) explaining the background to the subject and the model. Results were reported in various papers (Parton 1981, Harvey 1982, Harvey and Thomson 1984). A full description of the model at around 1985 is contained in Thomson (1987). Some illustrative results from its use at that stage give a flavour of its character and purpose.

Table 5.4 shows estimates of the three major effects of the CAP compared with free trade in 1984 (extracted from Harvey and Thomson 1984). These are:

a)   income gains to producers (in the first instance, farmers)
      through high CAP support prices;
b)   losses to users (consumers) of EC farm products from
      the same cause; and
c)   losses to taxpayers, who must finance the net cost of the CAP
      (after import levy and other minor revenues are taken into account).

The producer and user effects are calculated commodity by commodity for each country, and the three components can be added to give an overall 'income' estimate. The national 'budget and trade' effect - the sum of the taxpayer component and the additional cost or benefit of intra-EC trade in CAP products - represents the impact of the CAP on each country's balance of payments.

Table 5.4:   Benefits and Costs of the CAP vs. Free Trade, 1984

Effect	Producers	Users	Taxpayers	Overall	Budget and Trade
Country			- million ECU -		
Germany	11423	-11811	-5470	-5857	-3137
France	10925	-8216	-3-26	316	1764
Italy	8153	-6887	-2914	-1648	-1671
Netherlands	3639	-2261	-948	430	2508
Belgium/Luxembourg	1631	-1396	-704	-469	-317
United Kingdom	5994	-7826	-3811	-5643	-2717
Ireland	701	-502	-185	35	760
Denmark	1733	-770	-379	584	1751
Greece	2339	-2104	-367	-132	410
EC-10	46540	-42548	-17783	-13791	0

Source: Harvey and Thomson (1984).

The economic significance of the results, if not their precise magnitude, is too well known to need much discussion. Three aggregate-level features only will be picked out here. First, at EC level, the CAP in 1984 was estimated to add some 47 billion ECU to producer incomes (allowing for additional variable costs, but not for increases in 'fixed' costs such as rents), but to decrease consumer welfare by some 42 billion ECU. Since these very large amounts roughly cancel out (a reflection of the near self-sufficiency of the Community in CAP products), the budgetary cost of the CAP, at nearly 18 billion ECU, dominates the net overall welfare cost, put at some 14 billion ECU. Second, the national balances of the three component effects vary greatly amongst member states, the Netherlands, Denmark and Ireland showing positive overall effects due to their relatively large farm sectors, while Germany and the United Kingdom exhibit substantial overall losses, mainly due to their large budgetary contributions, accompanied in the UK case by demand-side losses much outweighing producer gains. Third, the budget and trade effects (which must net out to zero at EC level) show a major balance-of-payments gain from the CAP for France, but the opposite for Italy. All these features can of course be analysed per capita or where appropriate by individual commodity, since the model works from detailed calculation at commodity-market level. In addition, the model contains details of the budgetary components (import and producer levies, export refunds, intervention expenditures and MCAs), which can be checked against published EC accounts and analysed to reveal specific policy impacts.

The model continues to be used and developed, particularly at the Reading University Centre for Agricultural Strategy, within the EC 'Disharmonies' project (Commission 1988a), and in current work at both Newcastle and Aberdeen Universities.

REFERENCES

Bauer, S. (1988) Agricultural Sector Models: Retrospect and Prospect, Paper delivered to EAAE Seminar on Agricultural Sector Modelling, Bonn.

Buckwell, A.E. and Thomson, K.J. (1979) A Microeconomic Agricultural Supply Model, *J. Agric. Econ.*, 30 (1), pp 1-11.

Buckwell, A.E., Harvey, D.R., Thomson, K.J. and Parton, K.A. (1982) *The Costs of the Common Agricultural Policy*, Croom Helm, London.

Butterwick, M. and Neville-Rolfe, E. (1968) *Food, Farming and the Common Market*, Oxford University Press, Oxford.

Centre for International Economics (1988) *Macroeconomic Consequences of Farm Support Policies : Overview*, Canberra.

Commission of the European Communities (1988a) *Disharmonies in EC and US Agricultural Policy Measures*, Brussels.

Commission of the European Communities (1988b) *The Future of Rural Society*, COM (33)501, Brussels.

Fearne, A. (1989) A 'Satisficing' Model of CAP Decision-Making, *J. Agric. Econ.*, 40 (1), pp 71-81.

Fischer, G., Frohberg, K., Keyser, M.A. and Parikh, K.S. (1988) *Linked National Models: a Tool for International Food Policy Analysis*, Nijhoff, The Hague.

Glass, J.C. and McKillop, D.G. (1989) A Multi-Product Multi-Input Function Analysis of Northern Ireland Agriculture. *J. Agric. Econ.*, 40(1), pp 57-70.

Hagedorn, K. (1983) Reflections on the Methodology of Agricultural Policy Research, *Eur. Rev. Agric. Econ.*, 10(4), pp 303-323.

Harvey, D.R. (1982) National Interests and the CAP, *Food Policy*, 7(3), pp 174-190.

Harvey, D.R. and Thomson, K.J. (1984) *The 1984/85 CAP Debate: an Evaluation and Some Observations.* Discussion Paper 8, Department of Agricultural Economics and Food Marketing, University of Newcastle upon Tyne.

Hazell, P. and Scandizzo, S. (1986) *Mathematical Programming for Economic Analysis in Agriculture*, Macmillan, New York.

Heidhues, T. (1976) National Policy Decisions as an Adaptive Process, *Eur. Rev. Ag. Econ.*, 3(3), pp 349-389.

HM Treasury (1970) *Britain and the European Communities: an Economic Assessment*, HMSO, London.

Keyser, M.A. (1986) An Applied General Equilibrium Model with Price Rigidities, Staff Working Paper SOW-86-10, Centre for World Food Studies, Amsterdam.

Langworthy, M., Pearson, S. and Josling, T. (1981) Macroeconomic Influences on Future Agricultural Prices in the European Community, *Eur. Rev. Agric. Econ.*, 8(1), pp 5-26.

Mahé, L.P. and Moreddu, C. (1989) Analysis of CAP Trade Policy Changes. In: Tarditi, S., Thomson, K.J., Pierani, P., Croci-Angelini, E. (eds) *Agricultural Trade Liberalization and the European Community*, Oxford University Press, Oxford.

McClements, L.D. (1973) Some Aspects of Model Building. *J. Agric. Econ.*, 24(1), pp 103-120.

McFarquhar, A., Godley, W. and Silvey, D. (1977) The Cost of Food and Britain's Membership of the EEC. Ch. 3 in *Cambridge Economic Policy Review*, Gower Press, Aldershot, and later CEPR volumes.

Morris, C.N. (1980) The Common Agricultural Policy, *Fiscal Studies*, 1(2), pp 17-35.

Munk, K.J. (1984) A Model to Evaluate the Effects of Changes in the EC Agricultural Policy. In: Dubgaard, A., Grassmugg, B. and Munk, K.J. (eds) *Agricultural Data and Economic Analysis*, European Institute of Public Administration, Maastricht, and the Statens Jordbrugs-okonomiske Institut, Copenhagen

Nerlove, M. (1958) *Dynamics of Supply: Estimation of Farmers' Response to Price*, Johns Hopkins Press, Baltimore.

OECD (1987) *National Policies and Agricultural Trade*, Paris.

Parton, K.A. (1981) Levy Policies and EC Agricultural Surpluses, *Oxford Agr. Studies*, 10, pp 153-169.

Rollo, J.M.C. and Warwick, K.S. (1979) *The CAP and Resource Flows Among the EEC Member States*, GES Working Paper No. 27, London.

Tarditi, S., Thomson, K.J., Pierani, P. and Croci-Angelini, E. (eds.) (1989) *Agricultural Trade Liberalization and the European Community*, Oxford University Press, Oxford.

Thomson, K.J. (1982) Information Requirements for Decision-Making in Agricultural Policy. In: Hanf, C.-H. and Schiefer, G.W. (eds), *Decision and Information in Agribusiness*, KWV, Kiel.

Thomson, K.J. (1987) A Model of the Common Agricultural Policy, *J. Agric. Econ.*, 38(2), pp 193-210.

Thomson, K.J. and Buckwell, A.E. (1979) A Microeconomic Agricultural Supply Model, *J. Agric. Econ.*, 30 (1), pp 1-11.

Thomson, K.J. and Rayner, A.J. (1984) Quantitative Policy Modelling in Agricultural Economics. *J. Agric. Econ.*, 35 (2), pp 161-176.

Tyers, R. and Anderson, K. (1988) Liberalising OECD Agricultural Policies in the Uruguay Round: Effects on Trade and Welfare, *J. Agric. Econ.*, 39(2), pp 197-216.

Wallace, H. (1989) The Best is the Enemy of the 'Could': Bargaining in the European Community. In: Tarditi, S., Thomson, K.J., Pierani, P., Croci-Angelini, E. (eds) *Agricultural Trade Liberalization and the European Community*, Oxford University Press, Oxford.

Warley, T.K. (1967) *Agriculture: the Cost of Joining the Common Market*, PEP Report, London.

CHAPTER 6

THE CAP DECISION-MAKING PROCESS

Andrew Fearne

INTRODUCTION

As the EC has grown from six to twelve members, so the tailoring of a common agricultural policy to suit the needs and desires of all the member states has become increasingly difficult to achieve. This is largely due to the inherent problems of reaching a consensus between twelve partners with such disparate agricultural sectors and policy objectives, but is also related to the complicated decision-making process of the CAP which, like the policies it generates, has remained more or less unchanged over the past thirty years.

To understand the development of the CAP, the implementation of policy mechanisms and the array of policy instruments which have emerged, an appreciation of the decision-making process of the CAP is therefore required. This chapter identifies the range of factors which influence CAP decisions and examines the roles of the various organisations which have an input.

The first and major part of this chapter focuses on the annual price review, which has become the main process through which the policy comes under regular scrutiny and the levels of the various policy instruments are set. There then follows a section describing the routine management of the commodity sectors and a brief concluding section which highlights the increasingly important role of the Commission and a greater level of 'automaticity' with respect to policy adjustment in response to changing market circumstances

THE ANNUAL PRICE REVIEW

As the main device for the implementation of policy decisions, the annual price review constitutes the focal point for any analysis of CAP decision-making. The concept, as Thomson (1984) points out, owes much to the British, who during the entry negotiations of 1962 and 1967 insisted on the introduction of

101

an annual review procedure similar to that held in the United Kingdom. As such, the annual price review of the CAP 'permits a regular examination of the state of the Community's agricultural sector and gives member states an opportunity to voice their preoccupations' (Harris and Swinbank 1978).

The Basic Decision-Making Structure

There are essentially two stages in the annual review - the Commission's price/policy proposals and the Council of Ministers' decisions thereon. In the vast majority of cases, the Council cannot proceed without a proposal from the Commission. However, while the latter does have a 'permanent right and duty to initiate action' (Noel 1985), the Commission is often 'obliged' to adjust its proposals in accordance with the Council's line of thinking, largely in recognition of the political obligations of farm ministers.

Figure 6.1 summarises the basic structure of the decision-making process at the annual price review, from which the interaction between the various organisations and interest groups (national, institutional and commercial) is evident. The Committee of Permanent Representatives (COREPER) was originally created by the foreign ministers of the 'Six', to maintain a permanent link between member state civil services and that of the Community, with consultative rather than executive powers. By virtue of the subsequent creation of the Special Committee on Agriculture (SCA), similarly consisting of member state civil servants, COREPER is effectively excluded from considering any but the financial aspects of the CAP, with the SCA given the explicit task of advising the Council on the agricultural policy proposals put forward by the Commission.

The Economic and Social Committee (ESC) has a broader consultative role to play, as it represents the views of employers, trade unions, consumers, etc., within the Commission, and as such constitutes the 'in-house' lobby group for those non-agricultural sectors likely to be affected by changes in Community policy.

The 'external' pressure groups which have access to the decision-making process at a Community level include representatives from farm union organisations such as COPA (the Community's farm union organisation), COGECA (the Community's organisation for agricultural co-operatives) and CEJA (the Community's young farmers' organisation), consumer organisations such as BEUC (the European consumer organisation), trade associations, such as COCERAL (the Community's grain trade organisation) and farm industry's bodies, such as ASSILEC (the European dairy industry association), through which specific national interests are initially channelled. [1]

---

1   Member state pressure groups such as the National Farmers' Union (NFU) and the Diary Trade Federation (DTF) in the United Kingdom, also lobby agricultural ministers and members of the European Parliament at a national level, during the period of negotiations within the Council of Ministers.

Figure 6.1: The Basic Structure of the CAP Decision-Making Process

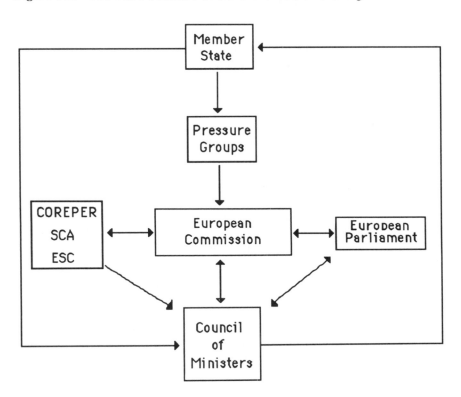

The European Parliament's role is somewhat ambiguous in that its views on policy initiatives are in no way binding, yet it has the power to dismiss the Commission and to reject the Council's proposals for the annual Community budget. The former requires a two-thirds majority vote and has never been effected, but the latter is often used in an attempt to frustrate proceedings and enforce the Parliament's opinion over the price package.

However, the role of many of these groups/organisations is largely symbolic, with most of the influential power being concentrated within a relatively small number of representative bodies. It is to the identification of these bodies and the extent of their influence over the decisions ultimately taken which we now turn, first by examining the price/policy proposal stage undertaken by the Commission, and second by looking at those factors which come into play when the Council approaches the task of reaching an agreement.

THE FORMULATION OF POLICY PROPOSALS: THE ROLE OF THE COMMISSION

As initiator of policy changes, it is the Commission's task to put forward a set of proposals on prices and related measures (non-price policy changes) which take account of all interested parties throughout the Community - in particular farmers, consumers and taxpayers alike. Thus, as the official 'manager' of Community affairs, the Commission is obliged to take account of the social, economic, financial and environmental implications of agricultural policy. The relative importance attached to these considerations varies from one year to another and is, to a large extent, dependent upon the economic and political environment in which the price review takes place.

Stage One - the Economic and Political Background

Every price review takes place against a particular political and economic background, which inevitably influences both the proposals put forward by the Commission and the subsequent decisions taken by the Council of Ministers.

Around October of each year, when the Commission starts to think about its price proposals, numerous organisations (agricultural ministries, farm unions, members of the trade, and so on) in all the member states collect and analyse information on estimated supplies and utilisation of the major agricultural commodities for the current marketing year, in order to make forecasts about the market outlook for the rest of the season. This provides the Commission with some indication of likely levels of intervention, prices, exports, and imports - from which it can estimate potential budgetary expenditure and determine the level of market management necessary. Thus, for example, the increase in milk output and the record level of butter production in 1982/83 highlighted the need for quotas, introduced in 1984. Similarly, the record cereal harvest of 1984/85 contributed to the pressure on the Commission to propose a 3 per cent reduction in cereal support prices in 1985/86.

More recently, the introduction of budgetary stabilisers has greatly increased the significance of supply estimates, as, for many commodities, price cuts follow automatically if the Maximum Guarantee Quantity (MGQ) is exceeded. Thus, for example, the estimate of the 1990 oilseed rape crop, some 30 per cent in excess of the MGQ, automatically resulted in support prices being cut by 14.5 per cent for the 1990/91 marketing year.

On the political side, national parliamentary elections and (to a lesser extent) local government elections regularly influence the course of events, if not for the price package as a whole then at least with regard to the particular arrangements for the member states concerned. For example, the presidential election in France in 1981 was a major factor in the agreement of the 1981/82 price review, reached two weeks before the election and representing an average price rise of over 11 per cent in national currencies. Similarly, the Italian coalition Government, on the verge of collapse in 1983, was granted substantial investment aids at the 1983/84 price review, at a time when a general election appeared imminent.

Conflict between member states can also substantially influence the level at which the Commission chooses to pitch its proposals. In 1982/83, for example, the Commission's desire to reform the wine regime was thwarted by a trade war between France and Italy, with Italian wine going sour under blockade on the French border.

Monetary developments also constitute a major input to the price review. The size and distribution of Monetary Compensatory Amounts (MCAs) across member states gives the Commission a clear indication of the level of common price rises necessary to facilitate a 'satisfactory' outcome in national currency terms. [2] The agreement in 1985 to create an agricultural ECU linked to the strongest currency, the Deutschmark, meant that, following the removal of the German positive MCA, the Commission would no longer be faced with the problem of having to increase common (ECU) prices simply in order to compensate farmers for reductions in national prices resulting from 'green' rate revaluations. This in turn enabled the weak-currency countries, particularly those outside the Exchange Rate Mechanism (ERM) of the European Monetary System (EMS), to take advantage of the inflated negative MCAs which resulted, and devalue their green exchange rates, as the UK did in 1989/90, securing an average increase in sterling support prices of 8 per cent when common prices for the EC as a whole had been frozen.

Stage Two - the Construction of a Price Package

The task of establishing a basis for the appropriate price adjustments for the various supported products is given to the Management Committees of the Directorate General for Agriculture (DGVI), who, having monitored the performance of the various sectors throughout the year, in conjunction with member state agricultural ministries, are normally in a position by October to give an early indication of the price adjustment deemed necessary. In addition to the Management Committees, the economics and external relations divisions are consulted, the former to assess the economic implications of alternative proposals and the latter for consideration of the impact of policy changes on third countries. During this preparatory period the Commission works in relative isolation, with little contact being made with the 'external' lobby groups identified above. Indeed, it is usually not until late October/early November that these groups begin seriously to enter the debate.

By the time any official consultation occurs, the broad outline of the Commission's price package has already been established. However, before the proposals are published (normally just before Christmas or early in the New

2   See Chapter 1 for a general description of the MCA system.

Year), considerable pressure is applied on the Commission from those parties with an interest in the outcome. Of these groups, COPA is by far the most influential. As the main representative of farmers' interests at the price review, COPA is always given ample opportunity to express its views. Indeed, the 'objective method', which takes account of cost increases and productivity gains in calculating the adjustment in farm prices necessary to keep farm incomes developing at the same pace as in the economy as a whole, has been adopted by COPA. It was, however, introduced by the Commission in 1973, and constituted a major input to the price review calculations until 1983, when Gundelach, the then Agricultural Commissioner, decided it was no longer appropriate to use the development of farm incomes as a basis for calculating changes in support prices.

BEUC, which represents the opposite end of the spectrum of interested parties, has a relatively small influence on the Commission's thinking. This is in part due to its lack of political strength, [3] but also to its overwhelmingly 'negative' stance towards the CAP. For example, at the 1984/85 price review, despite a reduction of around 6 per cent in average farm income in the Community over the previous year, BEUC was demanding a reduction in support prices of 8 per cent.

The arguments put forward by the various groups are relatively continuous from one year to the next: COPA consistently reminds the Commission of its duty to respect the basic objectives of the CAP and the fundamental principle of Community Preference. The objective of food security is often used as an 'excuse' for over-production, while the allowance of cheap third country imports is considered a threat to the stability of Community markets and producer returns (COPA 1981). BEUC, on the other hand, tends to concentrate on that clause of Article 39 of the Treaty of Rome which refers to food being supplied at reasonable prices (an objective rarely referred to and scarcely recognised by COPA) and emphasises the need for a better balance between consumption and production, favouring social and regional policies to price support as a means of supporting farm incomes.

Of the 'internal' influences, the European Parliament is generally supportive of COPA. This is perhaps not surprising given the importance of the rural vote in most of the European Parliamentary constituencies. However, as noted above, the opinion of the European Parliament (and its Agricultural Committee) carries relatively little weight, although the Commission cannot proceed with a proposal until the European Parliament has given its views. Indeed, it is this ability to hold up proceedings which the European Parliament generally exploits, particularly regarding budgetary issues, where its main executive powers are concentrated. For example, at the 1984/85 price review,

3   BEUC is an extremely small organisation, with little official or financial support from within the member states. Chapter 7 comments further on the relative weakness of the consumer voice in European agricultural policy decisions.

the European Parliament voted to withhold the budget rebates agreed for Germany and the United Kingdom until those two countries moved closer towards an agreement on the controversial price package.

The ESC as Neville-Rolfe (1984) observes, is rarely able to agree on a firm position to adopt, but when it does it also tends to err on the side of the Community's farmers. However, like the European Parliament, its influence over the outcome is limited and its opinion is generally reviewed as a formality rather than a major 'cog' in the decision-making machine.

The views of the SCA and COREPER do not really come into play during the proposal stages, but rather during the period of negotiation and consultation when agricultural ministers require advice on specific policy issues.

While the main purpose of the annual review is to adjust the level of price support, the importance of non-price policy adjustments has significantly increased since the Commission chose to take positive action to avoid the budgetary crisis towards which the Community was heading, via the CAP, from the late 1970s onwards. The growth of surplus production and associated budgetary costs, under a system of open-ended price support, has led the Commission to introduce a range of ancillary measures - MGQs, co-responsibility levies, production quotas and so on - the implementation and adjustment of which have become an integral part of the overall price package. In recent years, the discussion over these non-price policy issues has overshadowed the debate over changes in support prices, a feature which is perhaps indicative of an underlying shift in the emphasis of the CAP. For example, in the milk sector, where quotas have been established at a level far above Community self-sufficiency, any interest in the support price for butter and skim-milk powder has been exceeded by the importance of the quota for milk production. Similarly, in the beef sector, increases in suckler cow premiums have taken on considerably greater significance than the reductions in beef intervention prices, particularly in the UK, following the removal of the variable premium scheme in 1989.

In addition to its recommended price/policy adjustments, the Commission also proposes, when appropriate, changes in the representative ('green') exchange rates of member states and their associated MCAs, in an attempt to close the gap between the level of common (ECU) prices and those expressed in national currencies. These agrimonetary adjustments have become an increasingly important factor, particularly during the negotiating stages, when farm ministers are seeking to gain something 'extra' for their own farmers. Their influence on the final outcome is evident from Table 6.1, which shows the extent to which price changes in national currencies exceed those in ECUs, in terms of both the Commission's proposals and the Council's decisions. While member states retain sovereignty over these agrimonetary decisions and are thus not obliged to agree with the Commission's proposals thereon, reluctance to accept the proposed adjustment has usually been confined to those member states with positive MCAs, as their removal (or reduction) would result in the reduction of support prices in national currencies, unless offset by rises in 'consumer' prices.

Table 6.1: Price Changes in ECUs and National Currencies

Year	Council Decision		Commission Proposal	
	ECU	National Currencies	ECU	National Currencies
1968/69	0.0	0.0	0.0	0.0
1969/70	0.0	0.0	0.0	0.0
1970/71	0.0	1.5	0.0	1.5
1971/72	3.5	4.0	0.0	0.5
1972/73	8.0	9.8	6.5	8.3
1973/74	5.0	6.5	2.8	4.3
1974/75	13.5	19.9	11.2	17.6
1975/76	9.6	13.6	9.2	13.2
1976/77	7.5	11.4	7.5	11.4
1977/78	3.9	8.2	3.0	7.3
1978/79	2.1	8.5	2.0	8.4
1979/80	1.3	7.4	0.0	6.1
1980/81	4.8	10.5	2.5	8.2
1981/82	9.2	10.8	7.8	9.4
1982/83	10.4	12.1	8.4	10.1
1983/84	4.2	6.9	4.2	6.9
1984/85	-0.4	3.2	0.8	4.4
1985/86	0.0	1.3	-0.3	1.0
1986/87	-0.3	2.2	-0.1	0.9
1987/88	-0.2	2.6	-0.5	0.2
1988/89	0.0	1.6	0.0	0.3
1989/90	-0.1	1.3	-0.2	0.7
1990/91			1.1	0.2

Source: Commission (1980), Neville-Rolfe (1984), Agra-Europe (weekly).

When the Commission finally publishes its proposals for consideration by the Council of Ministers, it also provides estimates of the main economic and budgetary implications. Some observers [4] have been critical of the Commission's attempts to justify its proposals, which either underestimate the cost of the package (in the expectation that it will remain unaltered by the

4   For example, Thomson and Hubbard (1982), Fearne and Ritson (1987).

Council) or exaggerate the consequences of not adopting the proposals, in order to 'shock' the Council into accepting them as they stand. Nevertheless, the significance of the Commission's calculations lies in the attention which they draw to the impact which adjustments to the CAP have on the Community budget. In particular, they provide a benchmark against which the consequences of subsequent demands from agricultural ministers and pressure groups can be compared.

However, the attitude typically adopted by the Council of Ministers towards the Commission's budgetary/economic appraisal of its proposals suggests that this particular part of the Commission's contribution to the price review is accorded little weight during the process of negotiation and arbitration which follows. In 1981, for example, Josef Ertle, the German Agricultural Minister, summed up the view of most of his counterparts regarding the proposed price increases for 1981/82, when he said '... the European Community cannot afford it, but it will have to' (*Agra-Europe,* April 3, 1981). The factors shaping such an attitude become somewhat clearer when one considers the second stage in the annual decision-making process, when the Commission's proposals come under the scrutiny of the Community's agricultural ministers.

THE DERIVATION OF POLICY DECISIONS - THE ROLE OF THE COUNCIL

The fundamental difference between the proposal stage of the price review and the period of negotiations and eventual decision-taking is that the factors taken into account by the Commission in the preparation of its proposals are, in the main, of a social, economic, financial or environmental nature, whereas those influencing agricultural ministers are almost entirely political. Moreover, while it is a relatively simple task to identify a regular pattern behind the formulation of the Commission's proposals, the range of factors causing the Council to deviate from the Commission's recommendations is wide and often difficult to distinguish.

To a large extent, the price package which the Commission puts together is free from political bias. However, as the price review progresses into January/February and farm ministers become involved, the interests of the Community's ten million farmers tend to be given increasingly more weight than those from outside the agricultural sector, whose interests are generally reflected (albeit indirectly) in the Commission's concern over the budgetary implications of the CAP. The extent to which this occurs depends largely on the priorities and the political weight (voting power) of each member state, and as soon as the Commission's proposals are published the bargaining process begins.

Stage Three - Negotiation and Consultation

Between publication of the Commission's proposals and the first Council meeting, there is a period (the length of which varies from one year to the next) when lobby groups, government ministers, the European Parliament and numerous other bodies express their respective views on the Commission's price package. This prior consultation serves to narrow down the key issues and clear the way for farm ministers to reach an agreement by the beginning of May, the start of the marketing year for milk and most livestock products. [5]

In addition to national representation within the Council of Ministers, the Council President is obliged to hold meetings with both COPA and BEUC, and report on these to the rest of the Council. Perhaps a reflection of the changing emphasis of the CAP over time, BEUC was first accorded this right of consultation in 1980, when the United Kingdom Minister of Agriculture, John Silkin, was Council President, whereas COPA has enjoyed this privilege since the early 1970s.

On the more technical details, the Council seeks advice directly from the SCA (consisting of agricultural 'experts' from the member states). As with the Commission's proposals, though perhaps to an even greater extent, the attention given by the Council to pressure groups at the Community level is largely symbolic, with most of the influence from farm unions, consumer groups, and so on, being concentrated through national lobby channels. Thus, the relative importance of factors likely to affect the outcome depends very much on the strength of national lobby groups.

The French farm lobby in particular has been able to force the hand of its agricultural minister on several occasions. In 1983/84, for example, rioting French pig farmers forced the newly appointed Minister, Michel Rocard, to insist on the total dismantling of French pigmeat MCAs, which they argued were inhibiting their exports. And in the following year riots, kidnapping, and the spraying of foreign meat with diesel induced Rocard to set up a loan fund for farmers with chronic cash-flow problems, contrary to European Community regulations and for which the French Government was subsequently taken to the European Court.

National parliamentary elections in member states have already been cited as one of the many influential factors during the price review. However, the degree of influence on the final outcome is largely dependent on the country's bargaining power. Thus, for example, Ireland (one of the less influential member states) was able to secure the promise of a major farm aid package for small Irish dairy producers (a relatively minor concession) in 1982/83, on the eve of the general election in February. At the other extreme, the German farm minister, Ignaz Kiechle, fearful of the reaction of Bavarian farmers to a cut in cereal prices, stood firm on the eve of important local government elections, in 1985 and prevented a decision being taken on the proposed cut of 1.8 per cent.

---

5   The marketing year for most crop products begins on July 1.

Concessions of this kind have become more difficult to secure since the more widespread use of majority voting in the Council of Ministers in accordance with the Single European Act. For agricultural legislation to be adopted by 'qualified majority', 54 or more positive votes are required from the 76 available. The weighting procedure (Table 6.2) is constructed such that the larger member states are prevented from voting as a group to impose legislation on the smaller member states. Thus in recent years the use of the veto under the 'Gentleman's Agreement' has been replaced with a more concerted effort to reach a compromise over those issues which have threatened to affect adversely any one member state disproportionately. For example, the lengthy discussions over the budgetary stabiliser for cereals in 1988 resulted, not in the refusal of certain member states to adopt a punitive system of automatic price cuts, but a compromise resulting in a guarantee quantity (165 million tonnes) which was 5 million tonnes greater than the one proposed by the Commission.

Table 6.2: The Distribution of Votes within the Council of Ministers

Country	Number of Votes
UK, France, Germany, Italy	10
Spain	8
Belgium, Greece, Netherlands, Portugal	5
Denmark, Ireland	3
Luxembourg	2
Total Votes	76

Although the interests of member states in the various aspects of the CAP change over time, partly due to changes in the economic environment and partly as a result of changes in government, the positions adopted over the annual price review are relatively consistent from one year to the next. The net beneficiaries from the CAP tend to defend its basic principles and stress the need to maintain farm incomes; and the net contributors to the budget, though concerned about their farmers' welfare (or at least their votes), tend to concentrate on the market imbalance and the effect this has on Community finances and their contributions thereto.

However, there have been some notable exceptions which serve to illustrate the conflicting and often ambiguous problems with which member states are confronted. Germany, for example, has a relatively large food import requirement for some products, and as such should have a preference for lower food prices. However, it also has one of the strongest economies in the Community, which has resulted in the value of the Deutschmark rising consistently over the last decade, against most of the other member state currencies. This in turn has led to the existence of (often large) positive MCAs, which the Commission has constantly sought to eliminate. As already explained, the removal of positive MCAs requires the revaluation of the 'green' exchange rate and thus the reduction of support prices in that country's national currency. As a result, Germany has often been among those member states pushing for the highest possible common (ECU) price rise, in order to compensate for the effect of the associated agrimonetary adjustments.

A similar situation has, at times, confronted the United Kingdom, which has had to deal with a fluctuating currency, resulting in periods of high positive MCAs and periods of high negative MCAs, both of which have often conflicted with the national economic goals at the time. Moreover, successive British Governments have found it difficult, particularly in recent years, to reconcile the conflict between seeking to gain concessions over budgetary contributions on the one hand, and maintaining support for British farmers on the other. Indeed, British farm ministers have been criticised by their European counterparts on occasions when Britain has been seeking to reduce the general level of European Community price support, while insisting on the maintenance of specific British support measures, such as the beef variable premium.

France too, though heavily influenced by its agricultural sector, has occasionally been caught between the economic goal of countering inflation and the political necessity of acceding to the pressure from the farm lobby. The latter objective has been facilitated by the falling value of the franc, the resulting growth of negative MCAs and their subsequent removal via green rate devaluations, but this has exacerbated the inflationary cycle. A clear example of this conflict of interests arose during the 1983/84 price review: Agricultural Minister, Michel Rocard, wanted the French government to devalue the 'green' franc and remove the entire 6.6 per cent negative MCA to add to the proposed increase of 5.5 per cent in 'common' prices; but Finance Minister, Jacques Delors, was opposed to any devaluation that would increase farm prices by more than the general rate of inflation (10.3 per cent). In the event, a compromise was reached, with only part of the MCA being removed, giving an overall increase in French support prices of 9.4 per cent.

Stage Four - Denouement and Decision

As already noted, the discussions of the Council are confined to those proposals put forward by the Commission. One would therefore expect a strong relationship between the Commission's proposals and the Council's decisions thereon. Table 6.1 showed this to be the case, with the price changes implemented by the Council following a similar pattern to those prepared by the Commission. While this is of interest in itself (as it confirms the degree of interdependence between the Commission and the Council), it is perhaps more important to note the consistency with which the Council 'tops up' the Commission's proposed increase by one or two percentage points.

Given that the Commission itself proposes agrimonetary changes, it might be suspected that there would be little room for 'improvement' via further 'green' rate manipulations by the Council. However, it is possible for market rates to fluctuate from the moment the Commission's proposals are published to the time when decisions are taken, thus giving greater scope for 'green' rate changes at a later stage in the proceedings. This occurred during the 1982/83 price review when an apparent impasse had been reached in March 1982. There then followed a realignment of central rates within the European Monetary System (EMS), due to the devaluation of the Belgian franc and the Danish krone, aimed explicitly at facilitating a price settlement by adding between 3 per cent and 5 per cent to the price increases (in national currencies) proposed by the Commission. Indeed, one senior Commission official was reported to have said 'even the EMS is now being run to suit the CAP' (*Agra-Europe,* March 21, 1982).

A point often ignored, but noted by Thomson (1984), is the effect of 'green' rate changes between price reviews. These have an immediate impact within the member states concerned and often serve to offset decisions taken within the price review framework. Generally speaking, the Commission prefers not to allow 'green' rate changes during the marketing year, due to the disruptive impact which this can have on trade. For example, the devaluation of the Irish punt in October 1986 resulted in livestock being smuggled from Northern Ireland to Eire in order to profit from the higher prices pertaining south of the border. The position also varies between products and depends on whether existing MCAs are pre-fixed or variable. For example, cereals constitute a substantial proportion of input costs for livestock producers and, because of this, the Commission will often accept a mid-term 'green' rate devaluation for milk and livestock products, but not for cereals, as this would constitute an unforeseen increase in production costs. This was the main reason behind the British demand for a 6 per cent devaluation of 'green' rates for milk and livestock only, in October 1986.

A final point which one should note is the important (yet discontinuous) influence which individuals can have over the course of events. The Council President has a vital role in maintaining a dialogue with the Commission and encouraging a compromise between conflicting national interests. The fact that the presidency rotates every six months between the member states obviates any risk of hegemony; but presidential influence has often been instrumental in the resolution of conflict between member states over particular policy issues. For example, the roles played by President Mitterand and his Agricultural Minister, Michel Rocard, were central to the agreement over dairy quotas in 1984, and three years later President Mitterand was again a key figure at the Copenhagen Summit, when heads of state adopted the budgetary stabiliser package which, only days before, farm ministers had rejected.

ROUTINE MANAGEMENT OF COMMODITY SECTORS

While the basic level of policy instruments is generally agreed at the annual price review, the routine implementation of policies under the various commodity regimes is left to the Commission, in association with the respective government authorities in each member state. Thus, for example, in the UK, the Intervention Board for Agricultural Produce (IBAP) is responible for the intervention system and exports while HM Customs and Excise takes control of imports.

Most of the work regarding the day-to-day operation of commodity support policies requires considerable work of a technical nature. For example, export refunds, import levies and MCAs are calculated weekly, while the markets require constant monitoring to ensure the effective implementation of particular policies, such as intervention which is generally triggered when market prices fall below a predetermined level.

To help with this work, the Commission is aided by the Management Committees, of which one exists for each of the commodities supported by the CAP. These committees, consisting of up to five representatives drawn from the national ministries and related bodies of each member states, presided over by a representative of the Commission, meet regularly (weekly in the case of the major commodities) to discuss the operation of existing regulations (including the tendering arrangements for export refunds, the opening up of intervention and so on) and proposals put forward by the Commission. Voting proceeds along much the same lines as in the Council of Ministers and a position is adopted by qualified majority.

The views of the Management Committee are in no way binding. They can give either a positive opinion, a negative opinion or no opinion at all. Whatever the outcome, the Commission is entitled to proceed as proposed. However, to the extent that the representatives at the Management Committees reflect the views of their respective ministries, the Management Committee procedure provides a useful forum for identifying and narrowing down those areas which are most likely to cause problems at the later stages in the legislative process. It has been argued that, in its 'management' of the CAP mechanisms, the Commission possesses far more influence over the development of product markets than is usually appreciated. This issue, as it affects the food industries, is explored further in Chapter 11.

In addition to the Management Committees, there are numerous other committees which meet less frequently or on an *ad hoc* basis to assist in the fine-tuning of policies where expert advice is required. For example, the Agricultural Advisory Committee, of which there is one for each of the Commodities under the CAP, represents the vocational and economic interests (producers, processors and consumers) of the sector involved and is frequently consulted over technical changes to commodity regimes. Other specialist Committees include the Standing Committee on Agricultural Structures, the

Standing Committee on Agricultural Research, the Standing Committee on Seeds and Propagating Material, the Standing Veterinary Committee, the Standing Committee on Feedingstuffs, the Standing Committee on Plant Health and the Standing Committee on Zootechnics, to whom relevant draft legislation is submitted for an opinion in much the same way as the Management Committee procedure.

CONCLUSIONS

The development of the CAP from the post-war period to the early 1980s reflects the emergence of a wide range of contrasting and often conflicting pressures on the process of European integration. These pressures have reflected both the political constraints facing member states, in particular the ceding of sovereignty and the defence of national interests, and the economic consequences of adopting a common policy of open-ended price support in a sector which has responded beyond all expectations.

One consequence of the failure of Community decision-makers to get to grips with the problems of farm surpluses and budgetary expenditure under the CAP has been the gradual increase in the profile of the Commission with respect to the management of CAP policies and the introduction of automatic, rather than negotiated, changes in the level of policy instruments in response to changes in the market situation. Thus, for example, the introduction of Budgetary Stabilisers in 1988 removed, to a large extent, the room for manoeuvre at the annual price review, with price cuts automatically being implemented if production exceeds the negotiated threshold.

This gradual, but relentless shift in responsibility for policy decisions away from the Council of Ministers and towards the Commission is in line with the thinking behind the Single European Act and the rationalisation of Community decision-making post-1992, when the European Parliament is also to gain greater influence over the legislative process and majority voting in the Council of Ministers is to become the rule rather than the exception.

REFERENCES

Agra-Europe (weekly).
Commission of the European Communities (1980) *Reflections on the Common Agricultural Policy of the European Community,* Brussels.
COPA (1981) *General Position of COPA Concerning Common Farm Prices for the 1982/83 Marketing Year,* Pr (81) 23, Brussels.
Fearne, A. and Ritson, C. (1987) The CAP in 1995 - *a Delphi Survey of Expert Opinion,* Report No. 30, Department of Agricultural and Food Marketing, University of Newcastle upon Tyne.

Harris, S. and Swinbank, A. (1978) Price Fixing Under the CAP - Proposition and Decision, *Food Policy,* Vol. 3, No. 4, pp 256-271.

Neville-Rolfe, E. (1984) *The Politics of Agriculture in the European Community,* European Centre for Political Studies, Brussels.

Noel, E. (1985) *Working Together - the Institutions of the European Community,* European Commission, Luxembourg.

Thomson, K.J. (1984) Agricultural Prices Within the Common Agricultural Policy. In: Thomson, K.J. and Warren, R. (eds), *Price and Market Policies in European Agriculture,* Proceedings of the Sixth Symposium of the European Association of Agricultural Economists, 1983, University of Newcastle upon Tyne, pp 391-406.

Thomson, K.J. and Hubbard, L.J. (1982) *Evaluation of 1982/83 Price Proposals for the CAP,* Discussion Paper No. 1, Department of Agricultural Economics, University of Newcastle upon Tyne.

PART II

THE CAP AND THE COMMUNITY

CHAPTER 7

THE CAP AND THE CONSUMER

Christopher Ritson

INTRODUCTION

In view of the considerable media attention devoted to the impact of the Common Agricultural Policy on food prices, it is perhaps surprising that such a small proportion of the academic work on the CAP should have been directed specifically at the consumer interest. It is therefore instructive to begin by quoting Adam Smith:

'Consumption is the sole end and purpose of all production; and the interest of the producer ought to be attended to, only so far as it may be necessary for promoting that of the consumer. The maxim is so perfectly self-evident that it would be absurd to attempt to prove it. But in the mercantile system, the interest of the consumer is almost constantly sacrificed to that of the producer; and it seems to consider production, and not consumption, as the ultimate end and object of all industry and commerce.' (Adam Smith (1776) *The Wealth of Nations*)

Substitute 'Common Agricultural Policy' for 'mercantile system' and the author could be an agricultural economist writing 200 years later!

In this chapter, I first briefly consider the nature of the 'consumer lobby' relating to the CAP. Then I attempt to identify various aspects of consumer welfare which, in principle, might be affected by the CAP. Third, I consider the objectives of the CAP in the context of the consumer. Finally, the CAP market policy is considered relative to the various consumer interests.

THE CONSUMER VOICE
IN EUROPEAN AGRICULTURAL POLICY

It is generally accepted that the consumer voice in Europe, in so far as agriculture is concerned, is weak. It is difficult to cite one major example of a CAP decision which has been influenced predominantly by the consumer interest. One reason for this is that the consumer voice has typically been poorly articulated, and its arguments have been inconsistent and incoherent.

Elaborating on this, one can criticise the approach adopted by the consumer lobby with respect to the CAP in three specific ways. First, often it seems to be the wrong thing which is criticised. For example, much attention is devoted to 'food mountains', and in particular to the destruction of food, and also to budgetary expenditure. But, arguably, food surpluses are only symptoms - often rather minor symptoms - of the real consumer cost of the CAP. Just what this is I explore below, but it is worth giving one example of this particular problem here.

The issue which perhaps stimulates more extreme reactions than any other is the destruction of fruit and vegetables under the CAP's compensation system. It is not difficult to provoke outrage in an audience on this issue. Pictures of cauliflowers or lemons piled up, waiting to be destroyed, raise strong emotions - particularly when it is pointed out that 'they can only be destroyed if an EC Inspector certifies that the produce is of a sufficiently high quality to be destroyed'!

However, in my view, the destruction of 'surpluses' of fruit and vegetables can be justified rather more than many aspects of the CAP. Production of fruit and vegetables is highly seasonal and varies substantially from year to year. Demand can be quite inelastic. Produce cannot be stored (or at least is very costly to store). In these circumstances it does not make good sense to allow market prices to fall to ruinously low levels for producers in times of glut. It is quite sensible, from the point of view of *all* sections of society, if small quantities are removed from the market in such periods, and the most efficient course of action may well then be to destroy the produce; and the produce destined for destruction must be of an acceptable quality, or the EC would find itself compensating farmers for produce which would never have found a commercial market (and consequently would not have depressed prices).

Second, and following from the above, it seems that in its more popular manifestations the 'consumer view' of the CAP becomes obsessed with the idea that the policy represents some kind of great evil, which has an adverse effect on the lives of 'ordinary people' in every conceivable way. This view has particularly manifested itself in the context of current interest in diet and health, where the popular media view of the CAP seems to be that, not only does it 'exhort substantial excess expenditure out of consumers', but it is 'trying to poison them as well'! I shall return to this issue below.

Third, I think that the consumer interest in the CAP is typically defined far too broadly to be helpful - for example, the following is an extract from the terms of reference of a report published by the British National Consumer Council entitled *Consumers and the Common Agricultural Policy*: The CAP affects people 'primarily as consumers of food but also as consumers of the countryside and as taxpayers' (National Consumer Council 1988, Terms of Reference).

Thus, the so-called consumer lobby seems to be interested in the impact of the CAP on consumers, not just as consumers of food, but also as taxpayers and as 'consumers of the countryside'. In practice this means regarding the consumer interest in the CAP as comprising virtually every aspect of national welfare influenced by the CAP, other than the producer interest. The argument then becomes very diffuse. It is much more helpful, in a discussion of the CAP and the consumer, to restrict the issue to that of consumers as consumers of food. That is the approach taken in this book, where other chapters are concerned with aspects such as the CAP and the countryside, and taxpayer and budgetary aspects of the Policy.

THE CONSUMER INTEREST
IN EUROPEAN AGRICULTURAL POLICY

It is possible to identify four main aspects of (food) consumer welfare which are potentially influenced by the Common Agricultural Policy.

Food Prices and Expenditure

Most obviously, the policy influences the prices paid by consumers for individual food items and thus the proportion of total consumer expenditure devoted to food.

Food Availability

Consumers now expect a wide range of foodstuffs to be regularly available in the shops throughout the year.

Food Security

As consumers, we are interested in minimising the possibility that basic foodstuffs may suddenly become unavailable, or so expensive that a substantial proportion of the population cannot afford an adequate diet.

Food Quality

This is perhaps the most difficult of the four areas to define and it embraces a variety of aspects of consumer welfare. The provision of a range of qualities of individual food items should be covered by the issue of food availability. But there is also a need for consumers to be informed adequately concerning what they purchase, and it is usually accepted as a responsibility of government to legislate to control food safety, in terms of health and hygiene, and food additives. Finally, there is the question of the overall nutritional balance of the diet, and what have become known as the 'diseases of affluence'.

121

THE OBJECTIVES OF
THE COMMON AGRICULTURAL POLICY

Table 7.1 lists the objectives of the Common Agricultural Policy as specified in Article 39 of the Treaty of Rome. The right-hand column makes an informal estimate of the extent to which these objectives might be regarded as either producer-oriented or consumer-oriented.

Table 7.1: The Objectives of the Common Agricultural Policy

The objectives of the Common Agricultural Policy shall be:	C	P
(a) To increase agricultural productivity by developing technical progress and by ensuring the rational development of agricultural production and the optimum utilisation of the factors of production, particularly labour;	75%	25%
(b) thus to ensure thereby a fair standard of living for the agricultural population, particularly by the increasing of the individual earnings of persons engaged in agriculture;	0%	100%
(c) to stabilise markets;	40%	60%
(d) to guarantee regular supplies;	100%	0%
(e) to ensure reasonable prices in supplies to consumers.	100%	0%

Source: Article 39, Treaty of Rome (HMSO 1962).

In the case of the first objective, agricultural economic analysis shows that, typically, the characteristics of agricultural product markets are such that the major part of the benefit of productivity improvement will be passed on to consumers. This issue is discussed in more detail in Chapter 10 but I have allocated the first objective as being 75 per cent consumer- and only 25 per cent producer-oriented.

The second objective is, however, unambiguously a producer-oriented objective. The error made by the architects of the CAP, of course, was to believe that objective (b) could be achieved via objective (a). In practice the contrary has been case. The very success of European agriculture in improving its productivity has meant downward pressure on market prices. This, in turn, has made it more difficult to achieve a fair standard of living for the agricultural population - certainly when viewed in the context of rising incomes elsewhere in the economy.

It is generally accepted that the severe instability which characterises many agricultural markets, if left uncontrolled, is undesirable. Therefore looking at the objective of stability, the benefits are shared between producers and consumers. There are, however, theoretical arguments which suggests that, in certain circumstances, consumers might benefit from fluctuating food prices.

This argument, in my view, is rather weak (see Ritson 1985) but, because of it, I have allocated this objective 60-40 in favour of producers.

The objective of guaranteeing regular supplies is closely associated with what I have specified as 'food security' and 'food availability'. It is thus a consumer objective, as is, of course, the objective 'to ensure reasonable prices in supplies to consumers'.

Adding these figures up and dividing by five leads to the conclusion that the objectives of the Common Agricultural Policy are biased, by nearly two to one, in favour of consumers. It is therefore difficult to criticise the CAP from a consumer perspective, merely by looking at its objectives. To assess the Policy genuinely we must consider what it does, not what it states it would like to achieve.

THE CAP MARKET POLICY

In practice, the CAP consists predominantly of a complicated set of regulations which control the marketing of agricultural produce, mainly the prices at which produce is marketed at the wholesale level. A description of these mechanisms was given in Chapter 1 (illustrated in Figure 1.1), but typically the controls involve:

a) A tax on imports to prevent them selling below a predetermined price - a tax which varies so as always to bridge the gap between international prices and this predetermined (or minimum import) price.

b) Intervention in the market to prevent internal prices in Europe from falling beneath a floor level, usually set a little below the minimum import price.

c) Subsidies on exports to bridge the gap between the floor (intervention) level and the prices at which produce can be disposed of on world markets.

The upshot of all three of these mechanisms taken together is that prices of agricultural products within the European Community - and thus food prices to consumers - are kept above, usually substantially above, prevailing international prices. Let us now consider how these regulations, and some other aspects of the CAP, influence the four aspects of consumer welfare identified above.

FOOD PRICES AND CONSUMER EXPENDITURE

According to *Which* (April 1988) 'the CAP costs every man, woman and child in the EC £110 per annum as a consumer and £59 per annum as a taxpayer'. Four examples of food prices in the United Kingdom are given (see Table 7.2).

Table 7.2: Examples of Food Prices

	United Kingdom Wholesale Price	World Price
Butter 250 g pack	47p	19p
Beef (Topside) kg	178p	112p
Sugar kg	37p	12p
Rice (American Long Grain) kg	65p	38p

Source: *Which* (April 1988).

The four products listed in Table 7.2 have of course been selected because they are products for which, typically, the gap between UK wholesale prices and international prices has been rather high (although as far as butter is concerned the gap has often been much larger than that indicated in the table).

There have been other much more sophisticated calculations of the consumer costs of the CAP. Probably the most careful, and best known, are those that have been derived from the so-called 'Newcastle CAP Model' (see Chapter 5). For example, in his study for the Trade Policy Research Centre, Hubbard estimates that:

'...consumer expenditure on food in the United Kingdom in 1983 totalled £32,600 million. The consumer cost of the CAP, estimated at £3,890 million, effectively represents therefore an explicit tax on food at around 14 per cent.' (Hubbard, 1989).

This estimate substantiates the 'rule of thumb' that others have sometimes applied - that the CAP is equivalent to a Value Added Tax on food of about 15 per cent.

How do we know? Figure 7.1 represents a highly simplified version of the kind of model used in making these estimates. Conceptually the problem is quite easy; we simply revalue consumer expenditure on food products at alternative 'world prices' to obtain the consumer cost of the CAP. (We can also estimate the subsidy required to dispose of surplus production to obtain the taxpayer cost.)

In practice, of course, the calculation is complicated and presents difficult theoretical problems. Information on the level of consumption and production is not too difficult to come by and therefore does not provide too many problems. The 'EC price' also can be obtained. There is a problem that the price consumers pay is at retail level, whereas these calculations will usually be undertaken further up the marketing chain, to be comparable with international prices. It is also not certain that the marketing margin (the gap between the retail and wholesale prices) will be independent of the wholesale price so that the whole cost of higher wholesale prices will be borne by consumers - but this is the assumption which is usually made. It is also necessary to be cautious in using EC support prices as indicating the prices on markets, as these increasingly are different from internal market prices. Nevertheless, estimating the level of agricultural prices at the wholesale level is a manageable task.

Figure 7.1: Simplified Model for Estimating Consumer Cost of the CAP

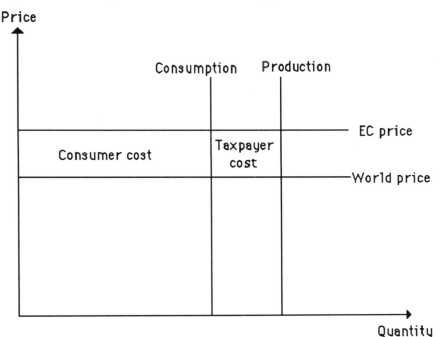

The real problem is the so-called 'world price'. In order to estimate the consumer cost of the CAP we must postulate an alternative scenario against which the present policy can be judged; and, irrespective of this alternative scenario, the so-called 'world price' is likely to be different from the prices currently prevailing on international markets. It is most unlikely to be correct simply to take the prevailing gap between European market prices and international prices, multiply this by the level of consumption, and call this the consumer cost of the CAP.

I have identified five possible alternatives, all of which might legitimately be used to measure the consumer cost of the CAP. These are listed in Table 7.3.

The most usual alternative scenario is to view the CAP against a situation in which no EC or member state policy influences agricultural markets. Under this alternative, less would be produced in Europe, more would be consumed, and this would have the affect of raising world prices. (This is the assumption that has been used in the Newcastle CAP Model in estimating the overall effect of the CAP on producers, consumers and the EC budget.)

Table 7.3: Alternative Scenarios to the Present CAP

(a)	Abolish agricultural policy.
(b)	Country considered withdraws from EC (or at least CAP).
(c)	Free trade, but sustain producer prices.
(d)	International movement towards free trade.
(e)	'Realistic' reform of the CAP.

The question then arises as to whether we are talking about the consumer cost of the CAP or simply the consumer cost of agricultural policy. Looking at the position from that of any individual member state, one can envisage the CAP no longer applying; but it is clearly quite unrealistic to regard the alternative as being no agricultural policy at all. Thus another scenario is the second in the table; that the cost of the CAP is viewed against the alternative policy that a country might be expected to adopt if it withdrew from the European Community (or at least from the Common Agricultural Policy).

A third possibility is to free trade between the EC and the rest of the world (that is to abolish levies on imports and subsidies on exports), but to sustain prices to producers by direct product subsidies. This is the scenario which the simple calculation of the current excess of internal prices over world prices most closely resembles. This is because, under scenario (c), producer prices (and thus the level of production) are sustained, and international prices would only be affected by the increase in EC consumption, as a consequence of allowing consumer prices in Europe to fall to world levels.

My fourth alternative scenario is described as 'international movement towards free trade'. Here I have in mind the kind of policy changes which are being negotiated in the context of the Uruguay round of the GATT, and in particular ideas relating to 'decoupling' of agricultural policy and price support (see Chapters 12, 14, 17). If the alternative is one in which prices are allowed to move towards world levels in the course of international agreement in which other major trading countries do likewise, we might expect international prices to rise substantially, thus leading to a lower estimate of the consumer cost of the CAP than that obtained under any of the other alternatives.

The fifth alternative is described as 'realistic reform of the CAP' by which is meant an alternative Common Agricultural Policy which might in some sense be regarded as politically feasible. This is the most difficult of the five scenarios to specify, but is perhaps the most likely in practice. (For a discussion of the reform of the CAP see, in particular, Chapter 16.)

Thus the consumer cost of the CAP (in terms of food prices and consumer expenditure) depends very much on the alternative scenario postulated. However, there is one point on which we can be reasonably certain. All five alternatives imply a net benefit to consumers. In other words, irrespective of which alternative scenario is used to judge the consumer cost of the CAP, all lead to the conclusion that there is a cost; but they all also lead to the conclusion that simply taking the prevailing gap between European market prices and world prices exaggerates that cost.

FOOD CHOICE AND AVAILABILITY

The main way in which the CAP influences the range of choice available to consumers within the European Community is by substituting European produce for imported produce. However, in the main, we are talking about agricultural commodities in which the nature and quality of the consumer product is not affected by the source of raw material supply. There may be some examples where consumers are adversely affected by, for example, having the range of - say - cheese or rice from non-European supplies restricted. But it is difficult to argue that most European consumers do not now have available to them a range of produce unparalleled in history.

It is mainly in the case of some fruits and vegetables that the CAP has a major influence on food availability (as opposed to food prices). Earlier, I defended the policy of destroying small quantities of certain fruits and vegetables as not being significantly against the consumer interest. Figures 7.2-7.5 show by way of example the proportion of produce withdrawn from the market for four products - tomatoes, cauliflowers, peaches and mandarins. In the case of cauliflowers and tomatoes, in general it does not seem that the proportion withdrawn is sufficiently high to be regarded as doing other than preventing prices from falling to very low levels; it is in the nature of the products and their markets, rather than the extremity of the policy, that these withdrawals occur. For most fruits and vegetables it is similarly the case that only relatively small quantities of produce are withdrawn (see Ritson and Swinbank 1988).

With peaches, and in particular mandarins, the position is somewhat different and a policy which leads to 15 per cent of peaches throughout the European Community (and 30 per cent in Greece) being withdrawn from the market must begin to look as if it *is* against the consumer interest; and with 80 per cent of Italian mandarins withdrawn from the market during the mid-1980s we clearly have a most peculiar market policy. Even here, however, this is not really significant for consumers. In general citrus fruits are plentiful, but this particular variety has little consumer appeal and the withdrawal from the market is concerned with budgetary expenditure and producer interests, not really denying consumers the benefit of the consumption of a valued product - though even mandarins were of value to consumers in 1988, when citrus produce was in short supply.

In addition, there is a reference price system, which attempts to establish minimum import prices for fruit and vegetables. Again, it seems to us that, in terms of its overall impact on average consumer prices, the system has only a modest impact. The intriguing feature about reference prices, however, in the

Figure 7.2: Percentage of Harvested Production Withdrawn - Tomatoes

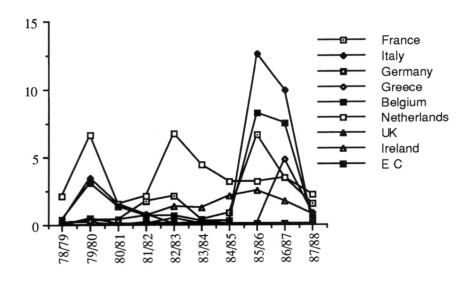

Figure 7.3: Percentage of Harvested Production Withdrawn - Cauliflowers

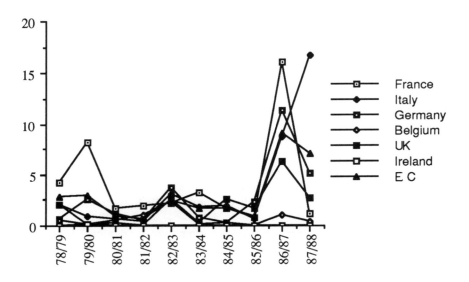

Figure 7.4: Percentage of Harvested Production Withdrawn - Peaches

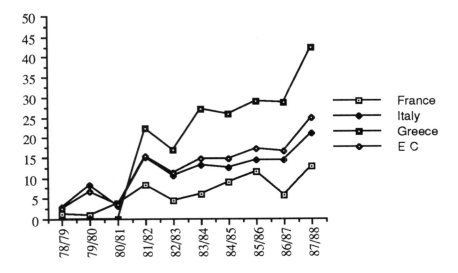

Figure 7.5: Percentage of Harvested Production Withdrawn - Mandarins

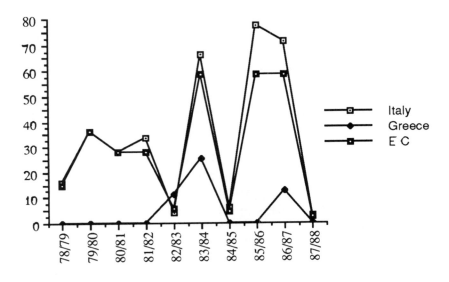

context of the present chapter, is the way the system seems to influence the availability and choice of food products. [1]

The minimum import price applied under the reference price system is different from that applying for most other agricultural commodities under the CAP, in two important respects. First whereas with most agricultural commodities, imports from third countries receive common treatment (that is, are subject to the same import levy) the reference price system is country-specific. Second, the system operates *ex post;* special import taxes (known as countervailing charges) are applied after produce from a particular third country has been observed to be selling on EC wholesale markets at less than the reference price (plus any import duty applicable).

Once countervailing prices have been applied the minimum price which must be respected is increased by the amount of the countervailing charge. The nature of competition on EC markets, or the inability of exporting countries to control the pricing of their produce, may make it impossible for suppliers to react to countervailing charges by raising their selling prices to the level required by the reference price system. In these circumstances countervailing charges can quickly rise to prohibitive levels until produce from the offending country is driven from the market. An example of this happening is given in Table 7.4.

From a consumer point of view the important aspect of this is that purchasers *are not* indifferent between the supplies of a particular fruit or vegetable coming from different countries - in the way, arguably, they are for most of the agricultural products covered by the CAP. Thus they may be denied access to varieties or qualities of produce during certain periods of the year because of the import controls exercised under the CAP - which are indifferent to variety or quality differences.

The support controls have a particularly interesting effect in the case of tomatoes and cucumbers (illustrated in Figures 7.6 and 7.7). The availability of good-quality cheap imported produce is seriously affected at a time when comparable EC produce is not available. Each year we find the same pattern applies - low prices in the first part of the year, when there are no import controls; produce from the major non-EC producers then suddenly becomes expensive at the time when the minimum import price comes into operation, and quite quickly produce from non-EC member states appears to be driven from the market.

The import duties charged on fruit and vegetables may also influence availability of supplies during certain periods of the year, as the duty concessions granted to supplying countries are often linked to seasonal calendars - that is, duties are highest during the EC production seasons. It has, for example, been argued by the Government of Cyprus, in its trading negotiations with the EC, that the date at which the full import duty on table

---

1   Chapter 15 considers reference prices from the perspective of the supplying country.

grapes becomes applicable to Cyprus is several weeks before there are plentiful supplies from EC countries. The consequence, it is argued, is that prices are depressed on the European market as third countries attempt to export as much as possible before the duty becomes applicable, and that there is then a period of relative scarcity until EC produce is available to replace the imports. Even then the varieties are different. Consumer choice *is* being restricted

Table 7.4: Impact of Countervailing Charges - Spanish Cucumbers (1984)

Date	Comment	Countervailing Charges Applicable to Spain
March	Average UK market price for cucumbers during March is 78 ECU/100kg	None
March 17	Low prices for Spanish cucumbers in UK and Germany lead EC Commission to apply countervailing charges on produce from Spain (including Canary Islands)	15 ECU/100kg
March 31	Countervailing charge raised	55 ECU/100kg
April 6	Countervailing charge raised again - now *exceeds* average wholesale price	83 ECU/100kg
April 14	Countervailing charge removed - no Spanish produce offered for sale on EC markets for several days	None

Source: Williams and Ritson (1987).

Thus there are instances in which consumer choice and availability are affected by the Common Agricultural Policy; but on balance the policy does not really come out too badly in this respect.

FOOD SECURITY

The Common Agricultural Policy influences the security of our food supplies principally by raising the level of self-sufficiency in individual agricultural commodities. The most important thing to recognise in this context is that food security and self-sufficiency in food supplies are not the same thing. Self-sufficiency is thought to contribute to food security, because the larger the

Figure 7.6: Tomato Prices in the UK, 1983

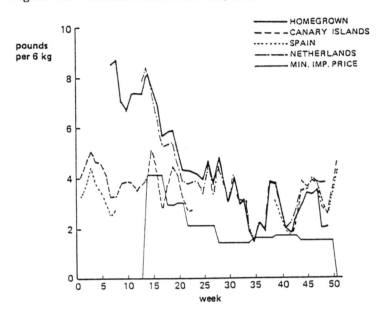

Figure 7.7: Cucumber Prices in the UK, 1984

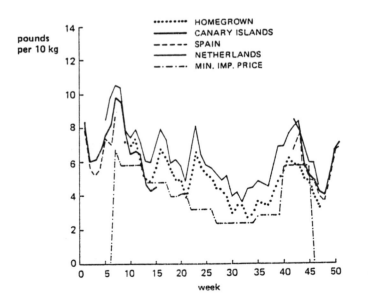

proportion of produce that comes from domestic supplies, the greater control the country apparently has over its food supplies.

There are two problems with this rather simplistic approach. The first is that there is insecurity attached to domestic supplies, as well as to imported supplies. In fact, typically, a greater degree of insecurity will emerge if a country is forced to rely solely on domestic production, because any individual country's production fluctuates proportionately more than the food supplies coming on to world markets. However, a country is able individually to use the world market to offset domestic fluctuations and so the potential availability of imports provides security against domestic fluctuations in food supplies.

Relying on imports to deal with domestic-orientated food insecurity is, however, only valid from the perspective of an individual country. If all countries choose to aim to be self-sufficient in food supplies, then there can be no international trade in agricultural products and the world trading system can no longer provide security against the failure of domestic harvest. Thus there is a paradox; if all countries aim to increase food security simultaneously by policies of self-sufficiency, they will collectively reduce security for all.

The second problem with equating self-sufficiency with food security is that there is insecurity associated with imported inputs and the use of non-renewal domestic resources, as well as with imported agricultural products. If a country boosts its self-sufficiency, but is dependent on imported inputs or is using up scarce domestic resources, it is reducing its food security in the longer term. Thus an agricultural policy directed genuinely towards food security would not be concerned with increasing the level of self-sufficiency using modern intensive methods of agricultural production, but be directed more towards the capacity of the farming system to sustain an adequate supply of food for the population in times of emergency using, perhaps, low-input techniques. [2]

Turning to the CAP, the verdict must be that it has increased self-sufficiency in Western Europe beyond the level that is necessary to attain food security - in the sense of ensuring the EC against over-reliance on imported supplies; and that it has tended to encourage a kind of agriculture which might not always be consistent with food security. Nevertheless, it would be difficult to argue that the policy had *decreased* our food security, though we could have achieved the same degree of security at much less cost in different ways. But as far as food consumers are concerned, the policy has perhaps had a mildly positive effect on this aspect on consumer welfare.

---

2   For a fuller discussion of agricultural self-sufficiency and food security, see Ritson (1980).

FOOD QUALITY AND NUTRITION

In general the CAP has a rather neutral impact on those aspects of 'food quality' concerned with consumer information and food safety. [3] It is, for example, sometimes argued that the classification systems used in the case of fruit and vegetables and carcass meats do not correspond very closely to genuine differences in consumer requirements; and it has been argued that, under the beef policy, high-quality meat goes into intervention, leaving poorer-quality beef on the European market. On the other hand, the EC does seem increasingly to be taking what might be argued to be a consumer-oriented approach in the case of things like pesticide residues and hormone implants. There is, however, a fundamental impact on this aspect of consumer welfare when considering the overall nutritional balance of the diet.

The CAP influences *patterns* of food consumption. The overall level of food consumption is reduced somewhat by higher food prices, but the main impact is derived from the fact that the policy raises the prices of certain products (relative to whatever alternative is applied) much more than others. Therefore it must have a significant impact on the balance of food products consumed.

Table 7.5 shows a 'league table' of the major food products according to the extent to which the CAP raises prices (and thus discourages consumption). Easily in first place is butter and other dairy products, where typically prices have been two or three times international levels. Next comes sugar, where world prices are usually very much below EC prices, though interspersed with short periods, every six or seven years, in which world prices rise, and have even exceeded EC levels. Next comes beef, where again estimates of EC prices are now usually well in excess of international levels.

The next group of products - essentially cereal products and the cereal-fed livestock - are those in which the CAP has a 'moderate' impact on food prices. In the case of vegetable oils and most fruits and vegetables (for different reasons), European consumer prices are not significantly above those which would apply in the absence of the Policy. Finally, there are one or two products where the Policy in effect subsidises prices to consumers. This has happened in the United Kingdom on occasions with lamb prices, where the variable premium (deficiency payments) system has sometimes meant that the market

---

3 Perversely, the reference price for fruit and vegetables may have a beneficial impact on food quality, as it is in the interest of non-EC supplying countries to attempt to avoid countervailing charges by concentrating on quality, and thus higher-priced, produce. One importer told us: '...the reference price system is bad in many ways, but it has had the effect of raising the quality of produce supplied to the market significantly over the past ten years' (Williams and Ritson 1987). In contrast he said the EC withdrawal system may be encouraging some EC producers to concentrate on the lower-quality produce.

price of lamb carcasses has been as low as half the price received by producers. The Policy has also attempted to aid the Italian tomato industry by subsidising processed tomato products very substantially.

Table 7.5: Impact of CAP on Food Prices ('League Table')

Butter (and other milk products)
Sugar
Beef

Bread
Pork
Poultry and eggs

Vegetables and oils
Most fruits and vegetables

Lamb
Canned tomatoes

NB: This hierarchy is broadly consistent with the rates of protection listed in Table 12.2.

Anyone who has followed the recent debate over the relationship between diet and health, and particularly the possible link between heart disease and diet, will not fail to notice that the Policy seems to have had the effect of discouraging, in the main, consumption of those products which medical experts tend now to regard as 'less healthy', and encouraging consumption of products more favoured from a health viewpoint. [4]

It is on this aspect that the popular conception of the CAP, as displayed in newspapers and television programmes, most blatantly misrepresents the consumer interest in the Policy. When the issue of diet and health is raised, typically what is presented will be pictures of fruit and vegetables being destroyed (thus 'the CAP destroys those foods which are thought healthy') and pictures of large stocks of dairy produce, sugar and beef (thus 'the CAP encourages the production of less healthy products'). There is a complete failure to distinguish between a policy which *encourages* the production of a product and a policy which *encourages* its consumption. In fact, by virtue of its reliance on price support, the CAP tends to discourage the consumption of those products of which it most encourages production. Coincidentally, it has succeeded in pushing the diet of European consumers in the direction that the

4   There is, of course, the associated question of the extent to which a government may be justified in manipulating food prices, to alter the diet of the nation, by adopting an implicit food policy. This issue is discussed in Ritson (1985) and Josling and Ritson (1986).

medical profession would now regard as 'more healthy'. For example, a recent study of the impact of the CAP on the consumption of food products in Greece shows a decline in consumption of sugar, meat and dairy products, and increases for citrus fruit, vegetable oils and vegetables (with no change for bread and cereals) - broadly consistent with the implications of Table 7.5 (Georgakopoulos 1990).

CONCLUSION

The verdict must be that the CAP has had an adverse impact on consumer welfare when viewed solely in terms of food prices and consumer expenditure. Simple estimates of the impact of the CAP on food prices will, however, almost certainly overestimate this. When viewed against a broader perspective of the consumer interest, the verdict is rather better. There are isolated examples of the policy adversely affecting the range of choice of food available, but on balance the policy cannot be said to have a major widespread impact in this respect. In the case of food security the policy has probably had a positive effect, although at much greater cost than could alternatively have been achieved, and at the cost of weakening the international trading system, which ultimately provides food security for all. Finally, although it has certainly never been any part of the intention of the Council of Ministers, the Policy, curiously, has probably had a positive impact on the overall nutritional quality of the European diet. Nevertheless, it remains the case that the CAP results in an implicit tax on food consumption, which has a regressive effect because lower-income households devote a higher proportion of their incomes to food than do the better off. Because of this most politicians argue that food taxes should not be contemplated; yet this is precisely the effect of the current CAP.

REFERENCES

Georgakopoulos, T.A. (1990) The Impact of Accession on Food Prices, Inflation and Food Consumption in Greece, *European Review of Agricultural Economics*, Vol. 17, No. 4.

HMSO (1962) *Treaty Establishing the European Economic Community, Rome 1957, Article 39.* London.

Hubbard, L.J. (1989) *Public Assistance to UK Agriculture* Report for the Trade Policy Research Centre (unpublished).

Josling, T.E. and Ritson, C. (1986) Food and the Nation. In: Ritson, C, Gofton, L. and McKenzie, J. (eds) *The Food Consumer,* Wiley, Chichester.

National Consumer Council (1988) *Consumers and the Common Agricultural Policy,* HMSO, London.

Ritson, C. (1980) *Self-Sufficiency and Food Security*, Centre for Agricultural Strategy, University of Reading.

Ritson, C. (1983) A Coherent Food and Nutrition Policy: the Ultimate Goal. In: Burns, J.A., McInerney, J. and Swinbank, A. (eds) *The Food Industry : Economics and Policies*, Heinemann, London.

Ritson, C. (1985) Some Observations on Price Instability, Agricultural Trade Policy and the Food Consumer. In: De Haen, H., Johnson, G. and Tangermann, S. (eds) *Agriculture and International Relations*, Macmillan, London.

Ritson, C. and Swinbank, A. (1988) EEC Fruit and Vegetables Policy in an International Context, *Agra-Europe*, Special Report, No. 32

Ritson, C. and Williams, H.E. (1987) Reference Prices and the Marketing Mix for Fruit and Vegetable,. *Food Marketing*, Vol. 3, No. 1. pp 61-76.

Smith, Adam (1776) *An Enquiry into the Nature and Causes of the Wealth of Nations*, Dove, London (First Edition).

Which (1988) Consumers Association.

Williams, H.E. and Ritson, C. (1987) *The Impact of the Reference Price System on the Marketing of Fruit and Vegetables in the UK*, Department of Agricultural and Food Marketing Report No. 31, University of Newcastle upon Tyne.

CHAPTER 8

THE CAP AND THE COUNTRYSIDE

Martin Whitby

INTRODUCTION

The countryside has been the focus of increasing contention in the last two decades. The vigour of the debate in the UK can be gauged from the flow and diversity of literature devoted to it; some of this has been polemical (Shoard 1980, Bowers and Cheshire 1983) whilst others have searched for a more analytical approach from a wide range of academic backgrounds (Williams 1973, Bell *et al.* 1978, Lowe *et al.* 1986). But the Governmental response, in terms of official reports published, legislation enacted and institutions established, has been even more notable. There have been examinations of the impact of farming on the soil (MAFF 1970), on the landscape (Westmacott and Worthington 1974), of the workings of the National Parks (Department of the Environment 1976; TRRU 1982), of the problems of the uplands (Countryside Commission 1983, 1984) and of agriculture and pollution (Royal Commission on Environmental Pollution 1979, Department of the Environment 1988), and innumerable local examinations of the interactions between those with competing interests in land. During the last four years Research Councils have diverted funds to investigating a wide range of countryside issues (Lowe *et al.* 1989, Harvey *et al.* 1990).

Several Acts of Parliament should also be mentioned in particular. The Countryside Act of 1968 extended and updated the more sonorously named National Parks and Access to the Countryside Act of 1949 and established the Countryside Commissions, of which the Commission for England and Wales has oversight of the National Parks. More recently, the Wildlife and Countryside Act (1981) attempted to regulate the impact of farming and forestry on natural environment and particularly on the countryside. This Act emerged from a vigorous parliamentary battle in which the conservation lobby pressed hard for direct controls on farming activities, but lost out in the end to the principle of 'voluntarism' which now prevails. Perhaps more striking have been the changes within agencies such as the Ministry of Agriculture, Fisheries and Food (in the Agriculture Act of 1986) and the Forestry Commission (in the Wildlife and Countryside (Amendment) Act of 1985), which have now had their objectives changed to include the conservation of the countryside and the promotion of its enjoyment by the public. This is a major change of emphasis for two departments which had hitherto pursued single-mindedly the goals of efficient production for the market.

In European terms it might be claimed that the UK was first to recognise the countryside and the rural environment as an element of land use policy which could be appropriately manipulated. However, the Netherlands has a highly sophisticated land use planning system (van Lier 1988) consistent with their major national reclamation efforts. Other member states have also responded to environmental land use problems in different ways (Dubgaard 1989, Chabason 1988). The European Community has demonstrated its awareness of environmental issues (Traill 1988) and has recently taken the first steps towards formulating proposals for the development of rural areas (European Commission 1988).

This chapter develops a definition of the countryside and then explores the impact of the CAP on the countryside over the past two decades, taking the UK as a case study for analysis. Finally, the recent attempt to retreat from the high level of farm protection in the CAP, by diverting policy initiatives to the development of the countryside and the protection of the environment, is explored.

WHAT IS THE COUNTRYSIDE?

Most people think of a dictionary definition of 'countryside'. The *Oxford English Dictionary* offers three related definitions:
1. A side of a country; now a favourite word of descriptive writers (1631).
2. The inhabitants of a tract of country (1840).
3. Rural (1853).

In each of these revealing definitions the date of the first recorded use in that sense is noted in brackets. The first suggests the origin of the word, and directs our attention to the landscape. The second emphasises the social connections of countryside - a concept which is still with us. The third is essentially its modern everyday use, including connotations of land use (for primary industry), population (thinly settled) and mystical attributes valued by many, including non-rural people (landscape, recreational options, open space).

The first noted use of 'countryside' as synonymous with 'rural' in 1853 was only two years after the Population Census had first recorded (in 1851) an excess of urban over rural population in this country. Britain was the first country in the world to find itself in that situation, which emphasises the long history of the concept of countryside. Thenceforward rural depopulation was a predominant trend until recent decades, when a somewhat uneven population 'turnaround' (Champion *et al.* 1987) has encouraged the view that there is no longer a depopulation problem in rural Britain.

In fact it is difficult to identify rural areas in any country with precision, and the United Kingdom is no exception. We have population data presented for Districts described as 'Remoter, Largely Rural' (Webber and Craig 1978) and we also have a completely different data set based on a cartographic identification and exclusion of urban areas (OPCS 1984). These two competing sets of information replaced an earlier system of identification based on types of local administration. For those who find the notion of a simple discontinuity between urban and rural areas, implied in all of the above approaches, inconvenient or unreal, there is also the work of Cloke and Edwards (1986), which led to the calculation of an 'Index of Rurality'. The latter ranks Districts on a scale of rurality, which runs from -12 to +12, from the Population Census.

Each of these data sets tells a different story and each offers answers to different questions. For the present descriptive task, the most useful data come from the cartographic analysis, which offers a broad range of information relating to a land use definition. Other workers have made use of the other sources above (for example, Whitby and Hubbard 1982, Champion *et al.* 1987).

The useful material which arises from the cartographic procedure relates to the rural population and its economic activities. In particular it tells us that there are some 5.3 million people living in rural Britain and that some 2.2 million of these were employed or self-employed in 1981. The distribution of employment in rural areas is shown in Table 8.1, where it is compared with that of the country as a whole. The notable feature of this display is the much greater importance of agriculture as an employer in rural areas (15.1 cf. 2.2 per cent). However, agriculture's share is less than several of the industrial aggregates displayed here. Most notable is that rural manufacturing exceeds it (19.5 cf. 15.1 per cent), but other sectors are also more important in rural areas.

Table 8.1: Distribution of Rural Employment by Sector, 1981

Sector	Number ('000)	Rural Employment % of Total	GB Employment % of Total
Agriculture	330	15.1	2.2
Energy and Water	60	2.5	3.1
Manufacturing	430	19.5	27.0
Construction	170	7.7	7.0
Distribution & Catering	400	18.2	19.2
Transport	100	4.7	6.5
Other Services	710	32.3	34.0

Source: OPCS (1984).

It is unfortunate that, because this data set has only recently appeared, and relates to only the 1981 Census, we are denied the opportunity of measuring the trends in the rural economy. This greatly weakens any attempt to evaluate specifically rural issues.

The discussion so far has ignored that other dimension of rurality, namely the land. Here the work of Best (1981) and others at Wye College has been particularly significant. The picture to emerge from such studies shows agriculture to be the main user of land, followed a long way behind by urban uses and forestry (Table 8.2).

Table 8.2: Land Use in the United Kingdom, 1971

	'000 ha	% of Total
Agriculture - of which:	18,831	78.2
Crops	7,227	30.0
Permanent Grass	4,926	20.5
Rough Grazings	6,678	27.7
Woodland	1,908	7.9
Urban Land	1,918	8.0
Other Land	1,436	5.9
TOTAL	24,093	100.0

Source: Best (1981).

However, tabulations such as this are somewhat limited in application because of their concentration on single land uses, which are treated as if they were all mutually exclusive. This follows simply from the way in which the data are collected, but it can be very misleading. For example, to say that we have 2.1 million hectares of land under forestry may be accurate, but it tells us nothing about the recreational use (actual or potential) which could be made of that area. Neither does it illuminate attributes such as landscape quality, water catchment yield or drainage. It is also completely uninformative as to scientific or conservation interest and does not record the distribution of ownership and tenure or the size structure of business units. Even where such data can be obtained from official sources, most of them are compiled on the basis of the single uses which are of interest to particular agencies.

To illuminate such issues we must turn to the data which are slowly beginning to appear from the computers of the conservation establishment. For example Briggs and Wyatt (1988) summarise sample data collected for the Countryside Commission by Hunting Surveys Limited, as shown in Figure 8.1. This material has been compiled from aerial and satellite photography and can therefore only tell us about vegetative cover as observed. It contrasts with the more traditional material, such as that reported by Best (1981), which is mainly derived from surveys and censuses of owners an occupiers. The advantage of such photographic data is that they are unambiguously tied to a spatial location, unlike the data from censuses of users, which are confused by changes in the spatial area owned or managed by the individual.

Figure 8.1: Land Use Cover in England and Wales - 1947 and 1980

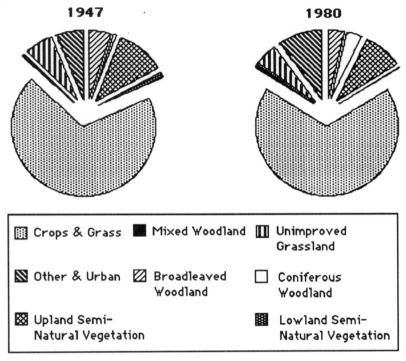

An important limitation of the data in both Table 8.2 and Figure 8.1 is that they present a static snapshot of land use at a point in time. They do not record the movements into and out of each category of use and consequently they reveal almost nothing about the dynamics of land use, which lead to a continual turnover of ownership and use patterns. This is important for predictive purposes because the offsetting elements of the change process are determined by different forces and would therefore require separate examination for forecasting their future course. Moreover the process of change is but poorly illuminated if we are only able to identify its net effect.

From the same source as Figure 8.1 it is possible to assemble a matrix of change, as in Table 8.3, which records all the identified movements between the categories shown. This presentation is a much reduced version of the original matrix, but it still reveals a great deal more than the stocks presented in Figure 8.1.

This presentation gives a picture of the movement of land areas between different cover types over this rather lengthy period. It is both descriptively more complete and a great deal more difficult to take in at a glance. The bottom row of the table summarises the total pattern of land use in 1980, telling us for example, that 64.9 per cent of the surface area was then under crops and grass. Following column f upwards until it intersects with the 1947 Crops and Grass row, we see that 59.6 per cent of the land surface was in the same category in both years, whilst the 1947 total was 65.4 per cent (from the 1947 total column). The 1980 total was achieved by adding 1.1 per cent from Broadleaved Woodland, 1.0 per cent from Semi-Natural Vegetation, 2.8 per cent from Unimproved Grasssland, and a remarkable 0.4 per cent from Urban and Other land, to the 1947 total, and deducting movements from Crops and Grass after 1947 to the various other categories, by 1980. Data such as these provide an opportunity to monitor changes in natural capital which are attributable to changing land use.

Table 8.3: A Matrix of Changing Land Cover in England and Wales:1947-1980

Cover in 1980 →	a	b	c	d	e	f	g	h	Total 1947
1947 ↓									
a Broadleaved Woodland	**3.3**	0.5	0.2	0	0	1.1	0.2	0.1	5.4
b Coniferous Woodland	0	**0.6**	0			0	0		0.6
c Mixed Woodland	0	0.1	**0.4**			0	0	0	0.5
d Upland Semi-Nat. Veg.	0	1.0	0	**8.3**	0	0.6	0.4	0	10.3
e Lowland Semi-Nat. Veg.	0.1	0.2	0.1	0.1	**0.3**	0.4	0.1	0	1.3
f Crops and Grass	0.4	0.3	0.1	0	0	**59.6**	1.6	3.4	65.4
g Unimproved Grassland	0.1	0.3	0	0		2.8	**2.5**	0.4	6.1
h Other and Urban	0	0		0		0.4	0	**5.4**	5.8
Total 1980	3.9	3.0	0.8	8.4	0.3	64.9	4.8	9.3	95.4

Source: Adapted from Countryside Commission (1990).

Note: Zeros represent small amounts, blanks indicate zero. After rounding errors the total percentage of surface area accounted for in the matrix is 95.4 in both years.

These movements into and out of particular categories are significant indicators of the turbulent pattern underlying the apparent stability of land use, in which the recorded total percentage shares in the two years are virtually unchanged. Other aspects of changing land cover which are notable from this table include the increase in the Urban and Other share from 5.8 to 9.3 per cent over this 23-year period. That total change would of course have been shown up by a simple static comparison: this presentation also shows that most of the urban increase came from crops and grass - that is, the better-quality land.

From the story so far presented, we have seen that the countryside covers 85-90 per cent of Britain's land surface and that it provides a residence for 9.25 per cent of the population. This means that the average density of population of the whole country is about 2.3 residents per ha, which is notably less than several countries (for example Japan, the Netherlands and Belgium). But simple arithmetic will also show us that the space per urban resident in Britain is only one-hundredth of that per rural resident.

Table 8.4: Population Density: Urban and Rural Britain (1981)

	Land Area ('000 ha)	Population ('000)	Population Density (residents/ha)
Total	22,998	52,760	2.29
Rural	20,990	5,339	0.25
Urban	2,000	47,421	23.71

Source: OPCS (1984), CSO (1988a) and Best (1981).

This qualifies the simple view of Britain as a 'crowded country': but it also points the contrast between urban and rural areas. So the answer to the question heading this section would be that the countryside consists of the rural landscape and the people who live and work in it. It provides a place of residence for some 5 million people and work for 2 million or so. Moreover it is highly valued by the rest of the population, many of whom visit it regularly. In economic terms, the countryside consists of a collection of mainly 'public goods' associated with land use. This rather austere definition may disappoint those who cling to a vision of the countryside as a collection of romantic images. Such visions may have their place (Mingay 1989), but their contribution to economic explanation must be mainly through the demand side of the countryside market. The supply of public goods in the countryside is usually taken to be a matter for the state to settle. This is mainly achieved through a range of public agencies whose role is to designate land for particular purposes (recreation, conservation, amenity and so on) and to manage that aspect of it. Designation is a complex, even subtle, process which seeks to ensure that the land concerned continues to be able to supply the good in question. The relationship between designation and the public good attributes of land has been examined in more detail elsewhere (Whitby 1990).

THE IMPACT OF THE CAP ON THE UK COUNTRYSIDE

Agriculture

It is appropriate to look first at the direct agricultural changes which have occurred during the period of British membership of the EC before exploring associated impacts on other countryside activities. In the simplest possible terms this can be done by reference to the Agricultural Census, which is collected each year from all agricultural holdings in the United Kingdom.

Britain joined the EC on January 1, 1973, which suggests a convenient starting date of June 1972. However, we must recognise the problem that the negotiations for entry to the EC continued over many years and the date of joining was known well in advance. So the data for 1972 will undoubtedly reflect an element of preparation for entry on the part of the more far-sighted farmers, as well as short-term fluctuations for individual items due to seasonal conditions.

Table 8.5 displays the main changes in agricultural land use in the United Kingdom over the 15 years following 1972. If we concentrate on the main magnitudes - the total cereals, permanent grass and rough grazings - the impression of rather slow change is difficult to resist. However within these totals there are some substantive shifts to be detected. For example:

• Although total cereals only increased by 3.6 per cent, there has been a major swing towards wheat and away from barley and oats. That small change also conceals a more substantial swing from spring to winter cereals.

• Within the broad 'other root' crop group, potatoes have continued their long slow decline in area, although it should be emphasised that total production was virtually the same in the two years, due to yield increases.

• The massive increase in oilseed rape production since 1973 is well known and is mainly attributable to the support offered under the CAP, although technological advance has played a part.

• Horticulture has declined in area, though area statistics may be particularly misleading in this case because of the variety of forms horticultural land use may take, from crops under glass to field-scale vegetables and orchards.

The net effect of these changes within the area of tillage shows an increase of 12.8 per cent, which is consistent with the much higher level of CAP cereal support prices compared with the level of guaranteed prices applying in the UK before 1974. There has been a reduction in temporary grass slightly larger than the increase in tillage. Permanent grass has increased, whilst the area of rough grazings has declined by more than 600 thousand hectares, mostly to forestry.

This cursory examination cannot be left unqualified. In particular it should be stressed that the reductions in area are not necessarily accompanied by reductions in output. Indeed in most cases the application of more inputs and the increasing specialisation of production units have more than offset any reductions in land area devoted to particular crops. Also, changes observable at the national level will not reveal some regional effects which may be more important in countryside terms. There are important statistical reclassifications occurring (for example, among the various types of grassland) which may also be reflected in these data, and the changes here are net, resulting from offsetting movements in different regions.

Table 8.5: Agricultural Land Use in the United Kingdom: 1972 and 1987

Item	1972	1987	1972-87 % Change
		Area ('000 ha)	
Total Cereals	3801	3940	+3.6
of which:			
wheat	1128	1992	+76.5
barley	2289	1836	-19.8
oats	315	100	-68.2
Total Other Crops	468	986	+108.5
of which:			
potatoes	237	178	-24.7
sugar beet	190	203	+7.1
oilseed rape	7	391	+5579.0
peas and beans	28	210	+640.7
hops	7	4	-41.9
Total Horticulture	273	197	-27.8
Total Tillage	4542	5123	+12.8
Temporary Grass	2359	1699	-28.0
Permanent Grass	4913	5108	+4.0
Rough Grazings	6619	5989	-9.5
Total Agricultural Area	18434	17919	-2.8

Source: UK Government (1974, 1988).

The changes in livestock numbers which have accompanied the land use shifts noted above are displayed in Table 8.6. The table contains only major categories of livestock which are likely to be observable in the countryside, with the exception of horses, which are not enumerated in the census but which have increased in number substantially over the period, and the groupings are broad.

It is notable that none of the products of these groups has declined, despite the reduction in number of breeding livestock associated with them. Pigmeat production has barely held its own in the face of higher cereal feed prices. Both milk and beef production are substantially greater than in 1972 and mutton and lamb have increased by more than 50 per cent. These increases in yield per animal and per hectare, which have been approached in many other EC member states, have been a major source of embarrassment to those trying to manage the CAP.

Table 8. 6: Main Livestock Numbers in the United Kingdom: 1972 and 1987

Item	1972	1987	% Change
		Livestock ('000s)	
Dairy Cows	3,325	3,036	-8.7
Beef Cows	1,476	1,339	-9.3
Ewes and Shearlings	13,106	18,065	+37.8
Sows and Gilts	960	813	-15.3

Source: UK Government (1974, 1988).

A further impact of the CAP which must also be examined has been borne by the farm labour force. The relevant data are summarised in Table 8.7. The distinction between full- and part-time workers is ignored in this presentation because it does not contradict the main theme, which is that hired workers have been decreasing in number despite the relatively high levels of unemployment (indicating slack labour demand) which have ruled throughout much of the period covered. A detailed econometric investigation carried out by MAFF officials (Lund *et al.* 1982) attempted to measure the relative importance of different factors in explaining this continuing decline in the number of hired farm workers. Their estimates show that the real decline in the prices farmers have received, since Britain joined the EC, will have accounted for only a few thousand of these jobs disappearing - quite a small share of the total decline experienced. Thus the CAP can apparently be credited with some impact on the rate of disappearance of agricultural jobs by slowing down the decline in the number of farm workers.

The impact of agricultural change on the countryside requires a more intimate local examination of data than is possible with the national aggregates in Tables 8.2-8.7. One approach to this would be through examination of the structural change in the industry, which indicates strong trends towards specialisation on the individual farm. These may be inferred from concentration ratios, reflecting the proportion of a particular activity found on the largest farms, or from observing the change in the number of producers of a particular output and the average size of their production units. The relevant data are summarised in Table 8.8.

Table 8.7: Changes in Agricultural Labour Use in the United Kingdom: 1972-1987

Category	1972	1987	% Change 1972-87
Farmers	297	287	-3.4
Hired Workers	306	306	-25.9
Total	710	593	-16.5

Source: UK Government (1974, 1988).

Table 8.8: Changes in the Size Structure of Enterprises in the United Kingdom: 1972-1987

Enterprise	Number of Producers			Average Size		
	1973	1987	change	1973	1987	Change
	(thousands)		%	(ha or number)		%
Crops/Grass	275.2	244.2	-11.3	43.9	49.2	+12.1
Cereals	125.1	89.7	-28.3	30.0	43.9	+46.5
Potatoes	64.6	32.8	-49.2	3.5	5.4	+55.4
Dairy Cows	94.0	50.1	-46.7	36	61	+69.4
Beef Cows	102.8	70.5	-31.4	16	19	+18.8
Breeding Sheep	85.3	84.7	-0.7	153	212	+38.6
Laying Fowls	99.3	42.1	-57.6	521	900	+72.7
Broilers	3.4	2.0	-41.2	16905	33000	+95.2

Source: UK Government (1974, 1988).

There is no need to examine this table in detail to see that in every case the number of holdings with each enterprise has declined and that the average size of units has increased, sometimes very substantially. This indicates a substantial increase in the concentration of production in the hands of the larger producers. For example, 70 per cent of dairy cows are now on holdings with 60 or more cows, 47 per cent of sheep are in flocks of 500 or more, nearly 70 per cent of laying fowl are in flocks of 20,000 or more... and so on (Annual Review of Agriculture 1988).

The concentration and specialisation which have occurred have required substantial investment in new technology and reorganisation of production units. This rate of investment was at a high level throughout much of the post-war period. However, during the period of UK membership of the EC a substantial change in the pattern of farming expenditure is notable. Even an industry which is not expanding has to replace its capital equipment in order to sustain production. The requirement for replacement is measured in terms of the annual depreciation of capital stock. If we wish to examine how expansive the mood of the industry is we should compare its rate of capital formation with its annual depreciation bill, as in Table 8.9.

Table 8.9: Income and Capital Formation in United Kingdom Agriculture: 1972-1974 and 1986-1988 (Current Prices)

	1972-4	1986-8
Farming Income	943	1374
Depreciation	267	1518
Gross Capital Formation (including work in progress)	608	917

Source: UK Government (1976),
          MAFF (1989).

The contrast between the two periods is notable. In the earlier period agriculture was forming new capital at a rate more than double its depreciation bill: more recently the position has been reversed and the industry is apparently disinvesting. The word 'apparently' is appropriate here because depreciation is a difficult entity to measure precisely for the individual firm and even more difficult for an industry with more than 100,000 firms. The comparison of Depreciation with Farming Income is relevant because investment must be funded out of present or future income. For a period an industry can finance investment by borrowing, as has certainly been the case in agriculture, but the long-term sustainability of such a process depends finally on the level of income available to service debts. Part of the decline in agricultural investment is attributed to the reduction in investment grants offered to farmers. During the 1980s three separate schemes have operated, each with successively lower percentage rates of grant offered for an increasingly environmentally benign set of investments.

This brief examination of agricultural change shows that, following EC entry, there have been several quite major changes in farm land use, in the level of output, in the resources tied up in agriculture and in the way in which they are distributed between farm businesses. The dominant themes of specialisation and technical efficiency are quite evident from these trends. But what is not evident, and will remain a matter for speculation, is the extent to which these changes are attributable to the CAP. That question can only be answered by reference to a detailed specification of the policies which would have alternatively been pursued, as previous chapters in this book have pointed out. Obviously some of these changes would have occurred under alternative policies: no doubt the farm supply industries would have increased their sales of technically superior inputs to farmers, no doubt many of the hired workers would have left, farm amalgamations and some concentration of production would all have occurred under most alternative policy scenarios. But, since we do not know the shape of this alternative situation, it is logically quite inappropriate to lay all of the perceived objections to what is happening in the countryside at the door of the CAP, as many critics have done (Shoard 1980, Bowers and Cheshire 1983). In the space available this theme cannot be fully developed; but it does provide an important qualification to conclusions we might wish to draw from the simple analysis of data presented and emphasises difficulties in comparing conclusions from different studies.

Forestry

In Wales, Scotland and the North of England, forestry provides an important alternative land use to agriculture, especially on land of poorer quality. Since much of the new forest planting occurring is coniferous and there have been questions raised as to its desirability on landscape and amenity grounds, we should briefly consider whether the CAP has had any impact on new forest planting. The Forestry Commission was established in 1919 and carries out both a supervisory role, with regard to private forests, and a production managerial role for the forest estates it owns. The current distribution of land uses between these two sectors, and the species planted are detailed in Table 8.10.

Table 8.10: Distribution of Ownership and Species in GB High Forest: 1987 ('000 ha)

Country	Commission		Private		Total	
	Conifer	Broad-leaved	Conifer	Broad-leaved	Conifer	Broad-leaved
England	198	41	198	385	396	426
Wales	127	6	49	54	176	60
Scotland	522	4	396	74	918	78
Total GB	847	51	643	513	1490	564

Source: Forestry Commission (undated).

After agriculture, forestry is the most important single land use in Britain. Moreover, it is continuing to increase its share of land cover under continued

Government policies of encouraging private plantation and subsidising the Commission's plantings. Since 1973, 392 thousand ha of new planting is recorded by the Forestry Commission and an increasing share of that has been in the private sector. The majority of new planting is coniferous and, in the interest of technical efficiency, there is a strong tendency to plant in larger blocks where possible.

Not surprisingly, land cover changes on such a scale have not passed without comment, and criticism reached a peak before Britain joined the EC. Some of this comment is now less appropriate than then, because the Forestry Commission is now paying much more attention to questions of landscape design. Its major problem is that mistakes in design may persist as a visual intrusion for half a century or more, whereas any improvements in design do not become fully evident for a decade or two after their introduction.

From the scale of planting since 1973 we can see that the CAP has apparently not generated sufficient farming wealth to check the process of land transfer from agriculture. But this comment is also subject to the qualification above regarding reference policies. What is clear from the workings of the land market is that policies designed to support farm incomes in the Less Favoured Areas (broadly upland Britain), where most of the new planting has occurred, have not been so attractive as to prevent the conversion of land to forestry. Alternatively we might say that the subsidies to encourage forestry must have been sufficiently attractive to bring about the transfer of land we observe. In so far as the policies with which they are competing derive from the CAP, it is therefore evident that a further 'hidden' cost of the CAP is the extra expenditure of taxpayers' and private money required to bring about the change. This picture of two different parts of the Government competing with each other to subsidise alternative activities on the same land, each therefore inflating its price, is not one that taxpayers can cheerfully contemplate.

Another aspect of increasing forestry which affects the quality of the countryside is its impact on rural employment. There has been debate elsewhere (Whitby 1989) about the extent to which forestry contributes to rural employment, and this is not the place to summarise that lengthy argument. However, it is worth noting that, although forestry does generate rural employment, it also removes any remaining agricultural employment on the land planted. Furthermore, because most of the actual employment in forests occurs at the end of the forest rotation, it will be several decades before the jobs generated by forestry actually materialise. Given that most improvements in forestry management since the war have reduced labour requirements at harvesting, new forest planting does not seem a very hopeful way of increasing employment in rural areas within the span of a forest rotation. This is not to deny that existing forests do employ some workers but to emphasise the long delay in the appearance of new jobs following new planting.

The discussion so far has concentrated mainly on the factors which determine what kind of countryside we have, that is, on the supply side of the countryside. The importance of the pattern of land ownership and occupation, the significance of the prices which drive farmers to respond in particular ways, the levels of employment in particular industries - all of these determine the form of countryside the rest of us might visit at the weekends or on holiday.

The impact of the CAP on these variables appears at first sight to be profound. It has apparently generated rapid increases in output and its impact on land values has further added to the cost of other land users including forestry and conservation interests. But the precise impact of the CAP on the countryside we see must be assessed in relation to the alternative policies that would have ruled. To assert that the CAP is the villain of the piece requires prior demonstration that the alternative policy scenario would have produced less damage to the countryside.

OTHER FACTORS IN COUNTRYSIDE DEVELOPMENT

To explain the development of the countryside in the UK solely in agricultural and forestry terms would be to miss crucial parts of the story. Two major elements must also be considered, namely the factors promoting demand for countryside goods and countryside policies and institutions which have emerged to regulate it.

Demand for Countryside Goods

The demand for countryside goods may be examined in terms of the changing components of consumer behaviour or in terms of the demand for public goods. This side of the market is less well understood by many, particularly those living in the countryside, and it is therefore essential to consider it here. Table 8.11 compares the development of various parts of consumer expenditure during the period 1973-87. During that period it shows that consumers' expenditure has risen, in real terms, by 32 per cent. Food and alcoholic drink have attracted a dwindling share of that expenditure, as measured at retail (in the case of food this also includes the mark-up in the catering sector). But agriculture produces only raw food, and its gross output grew over this period by some 17.2 per cent. Measured at the same prices, gross output was equivalent to 7.6 per cent of consumer expenditure in 1973 and 6.8 per cent in 1987. Its increasing absolute size compared with consumers' expenditure is consistent with the growth in self-sufficiency in food production from 62.3 per cent to 73 per cent over this same period. In comparison with the static growth of food we can see that expenditure on the array of goods associated with leisure and outdoor recreation have all grown very dramatically, increasing their share of consumers' expenditure. This group of expenditures together moved from 19 per cent to 22.8 per cent of total expenditure whilst food expenditure declined from 21.5 to 15.3 per cent.

Table 8.11: Expenditure on Leisure Related Goods: 1973-1987

	1973	1987	1987/1973
	(% of Total)		% Change
Real Consumers' Expenditure on:			
Food (including catering)	21.5	15.3	-5.7
Alcoholic Drink	7.9	6.3	+12.7
Travel - vehicle expenses	10.9	11.7	+42.9
- fares	3.6	3.4	+25.1
Recreation - TV, radio, video	1.2	2.5	+169.0
- sports and other equipt.	1.9	2.8	+91.8
Household Expenditure Abroad	1.4	2.4	+130.3
Total Consumers' Expenditure	100.0	100.0	+32.4

Source: Calculated from CSO (1984, 1988b).

Economists will recognise the familiar proposition underlying this evidence that the demand for food is **income inelastic** whilst that for leisure and associated goods is **income elastic.** The relationships for leisure expenditure have to be inferred indirectly because there are no direct estimates of the relevant elasticities for these activities in aggregate. Whilst there are many such measures for individual sites for outdoor recreation, we have no similar measure for aggregate outdoor recreation (see Walsh 1986, Willis *et al.* 1988). Meanwhile for food, the position is relatively clear, though even here most of the measures of income elasticity relate to food at retail: in general, we would expect the income elasticity of expenditure on food to be higher than that for farm products - because of the extent to which people demand more processed and convenience products as their incomes increase (rather than just more food). It is of course food which is more relevant when expressing different ways of using land. The importance of income elasticities is that they predict the rate of growth of expenditure on the goods in question as total incomes increase; the more elastically a good is demanded the more of the extra expenditure it will attract.

Another way of groping towards the relationship between income and recreational and environmental demands is to focus on the growth in participation in facilities over time. Such data are not collected routinely by any central authority, but the data in Table 8.12 have been collected from individual organisations.

The growth in the membership of these agencies gives some indication of the increasing enthusiasm for environmental and recreational matters in the countryside. This growth together with that in expenditure on leisure and recreational goods shows that the income elasticity of demand for what the countryside has to offer is a great deal higher than the demand for raw food. Another indicator of leisure growth is displayed in Table 8.13.

These data indicate only a slight decline in the length of the working week, over the period shown. However, the relative decline in manual work will have been accompanied by a reduction in the working week for all workers on average. By contrast annual holiday entitlement has grown substantially. The number of pensioners has increased, and their ability to enjoy leisure will have been enhanced by greater access to public and private transport. The number of unemployed has also grown steeply during this period. These are people with leisure time, although their scope for recreational participation may be severely constrained by lack of income. It is both intuitively obvious and consistent with economic theory that leisure time available has risen with incomes.

Table 8.12: Membership and Income of Countryside Organisations: 1973 and 1987

Organisation	Membership ('000)			Real Income (£'000)		
	1973	1987	% Change	1973	1987	% Change
National Trust	350	1,546	+340	24,015	87,083	+262.6
Ramblers Association	26	57	+119	178	532	+199.2
Council for the Protection of Rural England	24	32	+33	180	332	84.7
Royal Society for the Protection of Birds	129	500	+288	n.a.	11,000	n.a.

Source: Obtained directly from the organisations, 1988.
Incomes are expressed in 1987 prices.

The attempt to analyse countryside goods must also recognise the economic nature of these goods, which makes them particularly difficult to provide. A recent survey of these goods (Whitby 1990) draws attention to the fact that some are pure private goods from the consumption of which others may be excluded and in which there is no rivalry in consumption. Such goods would traditionally have included sports such as shooting and fishing which are sufficiently valuable to justify their owners excluding others from access to them. Others more recently added to the menu would include clay pigeon shoots and latterly the esoteric sport of bungy jumping. Exclusion from these activities is possible because they are carried out at a precise location, they require specific technology and access can be rigorously controlled. However, other countryside goods may be both non-rival and non-excludable which means

Table 8.13: The Growth of Leisure Time: 1973-1987

	1973	1987
Average Total Weekly Working Hours		
(full-time workers) Men - manual	46.7	44.5
- non-manual	38.8	38.6
Women - manual	39.9	39.5
- non manual	36.8	36.7
Annual Holiday Entitlement		
Modal Length - manual workers (weeks)	3	4+
Number of Pensioners (million)	8.3	9.6
Wholly Unemployed (million)	0.6	3.1

Source: *Employment Gazette* (1974 , 1978).

that their owner cannot be rewarded for providing them through markets and there is uncertainty as to the strength of demand for them. These are the classic countryside goods which, it is generally assumed, must be provided by the state. Such goods tend to be available across large areas and they include most of the environmental and scenic amenities of the countryside.

Typically state provision of countryside goods has been through a process of designation by governments. National Parks, Areas of Outstanding Natural Beauty and Green Belts as well as analogous areas in Scotland and Northern Ireland now account for a major share of the surface area of the UK. Because the designations are not all mutually exclusive (Green Belts may exist within a National Park, and so on) summing the total of these areas would involve double counting of actual space. That would make it difficult to interpret the resulting 10 million hectares in relation to the 20 million total area under agricultural and forestry use. The 10 million hectares does, however, represent the total extent of public commitment to the provision of such public countryside goods which has developed over the last few decades. It is also notable that some public goods - such as footpaths and bridleways - are not amenable to measurement in spatial terms.

Although there seems a considerable consensus that many countryside goods are pure public goods in practice, that view could be challenged in the light of the emerging market for such goods. In particular it is now the case that a number of wildlife groups are becoming significant landowners. Hodge (1990) reports the total areas owned and managed by seven such groups to be approaching half a million hectares. Major shares in this total are held by the National Trusts (for England and Wales and Scotland) but more recent landowners, such as the Royal Society for the Protection of Birds and the Royal Society for Nature Conservation and its affiliated Local Trusts, are now becoming significant landowners too.

The general conclusion from these leisure demand indicators would appear to be that they offer only a partial explanation of the increase in countryside recreation demand. The growth of income and the access this provides, mainly through mobility, would seem to be a more potent explanation of the trends we are examining. For these reasons it is expected that the demand for environmental goods from the countryside will continue to grow faster than the demand for raw food. This conclusion is particularly significant for those deriving their incomes from primary production in the countryside and there are already signs that this is being recognised by producers, as well as by the policy-makers who influence their activities.

CAP REFORM AND THE COUNTRYSIDE

The UK Government recognised the growth of environmental demands when it began explicitly to move towards more directly environmental policies for agriculture. Different observers will cite different turning-points, but an obvious candidate would be the statements which emerged in the summer and autumn of 1985 as the first major sign of a shift in the emphasis of policy. The 1985 Wildlife and Countryside (Amendment) Act introduced a conservation objective to the activities of the Forestry Commission and this was followed by the Agriculture Act of 1986, which explicitly assigns environmental and conservation responsibilities to MAFF.

The Agriculture Act of 1986 also introduced Environmentally Sensitive Areas (ESAs) within which farmers are offered incentives to farm in ways consistent with environmental conservation. Nineteen such areas, covering 0.8 million hectares, have now been designated and the MAFF is claiming a major success for these policies. How cost-effective they will prove remains to be seen, and indeed will be seen within the next three years because evaluations of these five-year schemes are required under the 1986 Act. Their impact on the conservation of the countryside will only become apparent over a much longer period. ESAs are an interesting example of the workings of the CAP, in that they allow the UK to exploit the EC Directive of 1985 on the Improvement of Agricultural Structures. The United Kingdom was the first member state to take advantage of this legislation, which may have been particularly appropriate as the UK proposed this part of the Directive.

A common view of such policies is that they reflect not so much a concern for the environment as a bid to reduce agricultural surpluses. However, a more accurate view of the MAFF position is that these policies are being pursued with environmental improvement as the main objective and the reduction of surpluses as a secondary objective. The extent to which such policies might contribute to the reduction of surpluses in the Norfolk Broads has recently been analysed by Colman and Lee (1987), who appraised the effectiveness of the Broads Grazing Marshes scheme. That scheme was essentially similar to, and has eventually been converted into, an ESA. Coleman and Lee estimated that the Broadland Scheme, which had been aimed at persuading farmers not to develop their grazing marshes for cereal production, had, over the three years for which it applied, reduced agricultural output by some 2 per cent per annum. That change in output will have reduced the amount of grain available for export refund, and hence the cost to British and European taxpayers.

The contribution of ESAs to improving the rural environment is bound to be difficult to assess, not least because the pay-off period from such policies will be long. Several years will be needed before the effects of such policies on ecosystems can be unambiguously demonstrated. Furthermore it is not easy to estimate the value of such changes which are not sold through markets and where one has to infer values from the way people behave towards them or attempt to extract other evidence from them by direct questionnaire techniques.

The dominant mood of those responsible for the CAP often appears to be one of some frustration. The price levels set within the policy are generally held responsible for the resulting surpluses and associated budgetary problems. The budget debate, which has dominated the 1980s, has reinforced the predictable suggestion that farmgate prices should be lowered. This debate has generated both speculation and concern in farming circles where the expected impact of such changes is viewed with alarm. Many arguments have been produced in defence of the status quo, and one of these has been that the CAP sustains fragile rural economies.

Given the small and decreasing share of farm employment in rural areas and the extent to which farmers buy their production inputs from urban suppliers, such an argument has a rather hollow ring. Nevertheless such views are heard in Brussels and they have recently resulted in the appearance of a communication from the European Commission (1988) entitled *The Future of Rural Society*. This report is of interest as an indicator of Eurocratic mood and it also demonstrates the enormous disparities between rural areas within the EC.

In attempting to cope with the massive variation within the EC, the report identifies three 'standard problems' of rural areas. These are:

a)   The pressure of modern development, which is particularly associated with urban fringe areas accessible to the towns, especially in Northern Europe.

b)   The problem of rural decline, which is mainly identified as a Mediterranean issue.

c)   The problem of depopulation and abandonment, where the policy aim must be to sustain a minimum population capable of protecting the environment.

These standard problems occur across the Community, but are concentrated in particular areas.. The first problem area is associated with strong urban links, whether by providing a residential location for commuting workers (in the South of England, East Anglia, the Po Valley, much of the Netherlands,

Flanders and Northern Germany) or a site for intensive tourist development, usually on the coast (Southern England and the Mediterranean coastal resorts).

The second problem, rural decline, is associated with a number of regions which have been losing population for decades due to migration (the Mezzogiorno, parts of Greece, Spain and Portugal, and Ireland and Northern Ireland). Many other rural areas are also in decline due to intra-regional migration mainly to urban areas. This problem is widely dispersed in outlying areas (North-West Spain, Western Ireland, Northern Ireland and Western Scotland as well as the Southern Mediterranean parts of the Community).

The solutions posed to these problems, although detailed, will resonate differentially in member states. Thus the first standard problem is to be ameliorated by measures to protect the environment, including land use planning. The popularity of this proposal amongst the UK farming community may be in doubt; but it might raise fewer arguments in other countries where planning systems are currently weakly developed.

The second standard problem, rural decline, is to be met by measures to improve farm structures and to promote economic diversification of the regions. The regional diversification approach mirrors the policies which have been applied unevenly across the UK for some decades. Regional development policies, and agencies to implement them, have waxed and waned, changed their areas of responsibility, been closed down and been revitalised since the 1960s. Amongst other member states, France appears to have adhered more consistently to given sets of regional institutions (Clout 1984) for promoting development. The key economic question here is what types of policy will best promote regional diversification?

The current pressure for reducing agricultural support might too easily create a policy vacuum to be filled by even less desirable policies. The notion of economic diversification has obvious strong attractions but instruments to use as a basis for such a policy are less obvious. For example, the provision of subsidised industrial premises has been the major single instrument for some decades. But, if that merely results in the transfer of urban jobs to rural areas, in what sense is society then better off? A more selective approach to diversification, seeking to attract industries which would have strong local linkages (for example, food processing industries) or for which there is increasing local demand (some services) would be one direction to test. However, at a time when unemployment is generally high it would complicate attempts to be selective in job creation if the meagre flow of potential factory tenants is to be assessed on the basis of local multiplier effects. Another, less fashionable, policy to promote diversification would be through regional payroll subsidies. Such a device also has the advantage that by subsidising **labour,** as opposed to **capital**, goods (that is, factories) it has a better chance of increasing employment.

The policies proposed for the third problem region include income and infrastructure support, measures to conserve natural capital of all kinds, and

the provision of information and advisory services. Under the heading 'Unremitting Effort' the report argues for:
- maintenance of rural populations at their existing level;
- small businesses should be reinforced where possible;
- national and Community intervention should be co-ordinated in programmes to ensure meeting the demand for services and infrastructure;
- conservation of the natural environment both for its own sake and as a base for tourist development;
- defence of the cultural heritage for similar reasons.

Even a modest achievement of most of these objectives would be difficult and costly and the report is weakest on how such a result might be obtained.

UK readers will notice the similarity between such policies and those operating within the Less Favoured Areas. An important difference might arise to the extent that such policies could be co-ordinated. Given the cost of pursuing such grandiose objectives across a broad front, policies will also have to be selective in application.

A few months after publication of the Report, the EC reformed its Structural Funds, setting five specific objectives for them:

a) developing less developed regions;
b) converting regions and smaller areas seriously affected by industrial decline;
c) combating long term unemployment;
d) combating youth unemployment;
e) developing (i) agricultural structures, and (ii) rural areas.

Objectives a, b and e(ii) apply to designated areas whereas the rest apply wherever appropriate conditions are met; for example, objective e(ii) applies in Less Favoured Areas and Environmentally Sensitive Areas. These designations are made by the European Commission after consultation with member states. The 'Objective e(ii)' regions of the UK include the Highlands and Islands Region, Western Dumfries and Galloway, most of Dyfed, Gwynedd and Powys, and parts of Devon and Cornwall.

What policies will emerge as a result of this action has been debated since publication of *The Future of Rural Society* (Commission 1988). The solutions posed to these problem areas are familiar in the United Kingdom. The response to modern development is to protect the environment; the response to rural decline is to promote development across a broad front; and the third (most intractable) problem will not be solved unless extensive farming systems are sustained, small businesses are strengthened and forestry activities built up, minimal services are supplied and the natural environment is conserved in the long run.

In conclusion the report points out that the countryside accounts for 80 per cent of the Community, in spatial terms (similar to the United Kingdom), but it also accounts for half of the population (compared with 10 per cent in the United Kingdom). This indicates the scale of the problem in Europe and emphasises the potential cost of such a policy across the Community, as well as a major political interest.

The EC report has also been subject to an extensive examination by the House of Lords Select Committee on the European Communities (1990). Their evidence, including submissions from nearly eighty separate agencies and individuals, gives a useful cross-section of policy positions on rural areas. The Committee presents considered opinions and a set of proposals too long to detail here. However, one proposal that is worth noting is the Committee's endorsement of the notion of 'one-stop shopping' for rural development advice. This idea has an EC label - the 'Carrefour' (translated variously as crossroads, intersection, meeting, symposium or circus!) where advice and information for rural development are consolidated and delivered by one agency. The defence of such a strategy is that participants in rural development are confused by the multiplicity of agencies and policies seeking to assist them. The Select Committee is forthright in rejecting the solution of establishing one single 'Ministry of Rural Affairs' but seems prepared to endorse the Carrefour proposal, partly on the grounds that it has already been tried and tested in the Peak District of Derbyshire in an integrated rural development scheme (Parker 1984). However, the Committee points out that the 100 Carrefours the Commission proposes to establish are unduly agrocentric: they should be 'entirely non-sectoral' in character. Whilst the Committee is unambiguous on that point it does concede elsewhere that the Agricultural Development and Advisory Service should become a 'Rural Development Advisory Service' to combine the existing functions of ADAS with those of the other rural development agencies *in the long run*. It would be most unfortunate if the fact of uniting all agencies under one umbrella with apparent agricultural origins deterred non-farming customers from making use of it. The reality of such a danger arises from the political isolation of MAFF in some rural areas, where it is bound to be seen as 'the farmers' friend'. The question is whether the tension of those traditional links can be relaxed sufficiently to allow conventional roles to be modified towards the wider community.

CONCLUSION

What is important is that our political machinery is increasingly recognising the rural environment as a legitimate matter of public concern. This recognition has been slow to materialise but will not disappear in the foreseeable future. The management of countryside environmental problems requires recognition of the relative role of supply and demand in generating them. In the case of the countryside, which is essentially a collection of environmental public goods, the arguments above suggest that demand is a much neglected side of the market and one to which increasing attention should be given. Indeed neglect of the growth of demand has been a major reason for our present problems of congestion and local intense overuse of the countryside.

The early EC moves towards a rural policy for Europe are notable not least in indicating the comparatively small and local scale of rural problems in the UK. Whilst the UK countryside is mainly in the first problem category, with comparatively small proportions of people living in the second and third problem areas, it seems likely that the latter two types of area will be more common in other member states.

The role for the UK in helping to mould emerging EC rural policies is potentially strong. There is a long tradition of both regional and environmental protection policies in the UK, which is becoming increasingly relevant in other member states. The challenge for the UK, and for the rest of the EC, will be to produce rural policies of comparable relevance and appeal across its huge and diverse rural areas.

REFERENCES

Bell, C., Newby, H., Rose, D., and Saunders, P. (1978) *Property Paternalism and Power*, Hutchinson, London.

Best, R.H. (1981) *Land Use and Living Space*, Methuen, London.

Bowers, J.K. and Cheshire, P. (1983) *Agriculture, the Countryside and Land Use: an Economic Critique*, Methuen, London.

Briggs, D. and Wyatt, B. (1988) Rural Land Use Change in Europe. In: Whitby, M.C. and Ollerenshaw, J. (eds) *Land Use and the European Environment*, Belhaven, London.

Chabason, L. (1988) The French Land Use Planning System. In: Whitby, M.C., and Ollerenshaw, J. (eds) *Land Use and the European Environment*, Belhaven, London.

Champion, A.G., Green, A.E., Owen, D.W., Ellin, D.J. and Coombes, M.G., (1987) *Changing Places: Britain's Demographic, Economic and Social Complexion*, Arnold, London.

Cloke, P and Edwards, G. (1986) Rurality in England and Wales 1981: a replication of the 1971 Index, *Regional Studies* Vol. 20 No.4, pp 289-306.

Clout, H. (1984) *A Rural Policy for the EEC?* Methuen, London.

Colman, D., and Lee, N. (1987) *An Evaluation of the Broads Grazing Marshes Conservation Scheme 1985-88*, Final Report submitted to the Countryside Commission, Department of Agricultural Economics, University of Manchester.

Commission of the European Community (1988) *The Future of Rural Society*, COM (88) 371, Brussels.

Countryside Commission (1983) *The Changing Uplands*, CCP 153.

Countryside Commission (1984) *A Better Future for the Uplands*, CCP 162.

CSO (Central Statistical Office) (1988a) *Annual Abstract of Statistics*, HMSO, London.

CSO (Central Statistical Office) (1984, 1988b) *National Income and Expenditure*, HMSO, London.

Department of the Environment (1976) *Report of the National Parks Policy Review Committee*, HMSO, London.

Department of the Environment (1988) *The Nitrate Issue*, HMSO, London.

Dubgaard, A. (1989) The Need for a Common Environmental Policy for Agriculture. In: Dubgaard, A, and Hjortsoj Nielson, A. (eds) *Economic Aspects of Environmental Regulations in Agriculture*, Wissenschaftsverlag Vauk, Kiel.

European Commission (Commission of the European Communities) (1988) *The Future of Rural Society*, COM(88) 371, Brussels.

Forestry Commission (undated) *Forestry Facts and Figures, 1986-1987*, Forestry Commission, Edinburgh.

Harvey, D.R. and Whitby, M.C. (1987) Issues and Policies. In: Whitby, M.C. and Ollerenshaw, J. (eds) (1988), *Land Use and the European Environment*, Belhaven, London.

Harvey, D.R., Whitby, M.C., Willis, K.J., Allanson, P., Adger, N. and Garrod, G. (1990) *Markets, Policies and the Countryside Change*, ESRC Countryside Change Initiative Working Paper 3.

Hodge, I.D. (1990) Land Use by Design? In: Britton, D.K. (ed.) *Agriculture in Britain: Changing Pressures and Policies*, CAB International, Wallingford.

Lowe, P., Cox, G., MacEwen, M., O'Riordan, T. and Winter, M. (1986) *Countryside Conflicts; the Politics of Farming Forestry and Conservation*, Temple Smith/Gower, London.

Lowe, P., Marsden, T., Munton, R., Harvey, D.R., Whitby, M.C. and Willis, K.G. (1989) *The Countryside in Question: a Research Strategy*, ESRC Countryside Change Initiative Working Paper 1.

Lund, P.G., Morris, T.G., Temple, J.D. and Watson, J.M. (1982) *Wages and Employment in Agriculture: England and Wales 1960-1980*, MAFF, London.

MAFF (Agricultural Advisory Council) (1970) *Modern Farming and the Soil*, HMSO, London.

MAFF (1989) *Agriculture in the United Kingdom - 1988*, HMSO, London.

Mingay, G.E. (ed.) (1989) *Rural Idylls*, Routledge, London.

OPCS (1984) *Key Statistics for Urban Areas: Great Britain: Cities and Towns*, HMSO, London.

Parker, K. (1984) *A Tale of Two Cities*, Peak Park Joint Planning Board, Bakewell.

Royal Commission on Environmental Pollution (1979) *Agriculture and Pollution, Seventh Report*, HMSO, London.

Select Committee on the European Communities (House of Lords) (1990) *The Future of Rural Society*, HMSO, London.

Shoard, M. (1980) *The Theft of the Countryside*, Temple Smith, London.

Traill, B. (1988) The Rural Environment: What Role for Europe? In: Whitby, M.C. and Ollerenshaw, J. (eds) *Land Use and the Rural Environment*, Belhaven, London.

Tourism and Recreation Research Unit (TRRU) (1982) *The Economy of Rural Communities in the National Parks of England and Wales*, Edinburgh.

UK Government (1974) *Annual Review of Agriculture*, HMSO, London.

UK Government (1988) *Annual Review of Agriculture*, HMSO, London.

van Lier, H.N. (1988) Land Use Planning on its Way to Environmental Planning. In: Whitby, M.C. and Ollerenshaw, J. (eds) *Land Use and the European Environment*, Belhaven, London.

Walsh, R.G. (1986) *Recreation Economic Decisions: Comparing Benefits and Costs*, Venture Publishing, Colorado.

Webber, R. and Craig, J. (1978) *Socio-Economic Classification of Local Authority Areas*, Studies on Medical and Population Subjects, No.35, Office of Population Censuses and Surveys, HMSO, London.

Westmacott, R. and Worthington, T. (1974) *New Agricultural Landscapes*, Countryside Commission, Cheltenham.

Whitby, M.C. (1989) The Social Implications for Rural Populations of New Forest Planting. In: Adamson J. (ed.) *Cumbrian Woodlands: Past, Present and Future*, ITE Symposium 25, Grange-over-Sands, HMSO, London.

Whitby, M.C. (1990) Multiple Land Use and the Market for Countryside Goods, *Journal of the Royal Agricultural Society of England.*, Vol. 151 pp 32-43.

Whitby, M.C. and Hubbard, L.J. (1982) *The Urban-Rural Income Gradient*, Discussion Paper 7, Department of Agricultural Economics, University of Newcastle upon Tyne.

Williams, R. (1973) *The Country and the City*, Chatto and Windus, London.

Willis, K.G., Benson, J.F. and Whitby, M.C. (1988) *Values of User Benefits of Forest Recreation and Wildlife*, Report to the Forestry Commission.

CHAPTER 9

THE CAP AND INTRA-EC TRADE

Caroline Saunders

INTRODUCTION

The EC was established on the principle of a common market between member states, as set out in Article 2 of the Treaty of Rome, which stated:

'The Community shall have as its task, by establishing a common market and progressively approximating the economic policies of member states, to promote throughout the Community a harmonious development of economic activities...'

A common market implies, not only the free movement of goods and services, as in a customs union, but also the free movement of capital and labour and the elimination of any factors inhibiting competition between member states, such as discriminatory taxes and state aids. How far the expectations of the architects of the EC have been met is questionable. Historically the EC has attempted to achieve the free movement of goods and services as was intended under Article 9 of the Treaty of Rome which states that the Community:

'...shall be based upon a customs union which shall cover all trade in goods and which shall involve the prohibition between member states of customs duties on imports and exports and of all charges having equivalent effect, and the adoption of a common customs tariff in their relations with third countries.. '

Therefore in analysing the impact of the EC on trade it is reasonable to assume that it was, and still is, in practice a customs union. The wider objective of the Community to establish a common market is the fundamental objective of the Single European Act 1987 given in Article 8A, which states that:

'The internal market shall comprise an area without internal frontiers in which the free movement of goods, persons, services and capital is ensured in accordance with the provisions of the Treaty.'

It is proposed that by the end of 1992 all barriers impeding the free movement of factors of production will be removed in addition to any remaining barriers to movement of goods and services. In achieving this the Commission detailed in its 1985 White Paper (European Commission 1985) 300 proposals and a detailed timetable for reform. These proposals were summarised by Lord Cockfield, the responsible commissioner, into three types of barriers to the existence of a common market - physical, technical and fiscal. Physical barriers are defined as frontier controls; technical barriers as non-tariff barriers, such as health and safety regulations; and fiscal barriers as differing taxes between member states.

In the case of agriculture, the expected impact on trade of establishing the Community differed, due to the adoption of the CAP. The CAP went beyond the aims of a customs union - free trade between member states and a common external barrier on trade with third countries - by establishing common internal support prices. However, the development of MCAs brings into question the extent to which the CAP has achieved either free intra-Community trade or common support prices. The impact of these policies upon trade within the Community is discussed in the following section.

TRADE THEORY AND THE IMPACT OF GOVERNMENT RESTRICTIONS ON TRADE

The theory of trade provides explanations of why trade occurs, its commodity composition, its direction and the terms at which this trade takes place. Trade theory illustrates that countries benefit from trade by obtaining either factors or products, which they can not produce, and/or through specialisation of production in those commodities in which they have comparative advantage. [1] Thereby they can consume a combination of commodities which they otherwise would not be able to achieve. Allowing free trade between countries therefore, in one sense, maximises world welfare and output. [2]

The conclusion that free trade 'maximises world welfare' implies that any government intervention in trade, in the form of tariffs or subsidies, or by other means, [3] will reduce welfare and output. For example, if a government imposes a tariff on imports, this will increase the price of the imported commodities relative to the exported commodities and domestic production. The production in the importing country will therefore shift from the

---

1   That is, commodities they can produce relatively cheaper.

2   The fact that countries specialise in those commodities in which they have a comparative advantage implies that world output is maximised relative to available resources. The conclusion that this imples maximum 'world welfare' is subject to a number of qualifications.   The most important of these are: a) that the argument is 'static', and does not take account of the fact that some trade restriction might stimulate growth; and b) that it takes no account of (or 'accepts') the distribution of income brought about by free trade.

3   The other means of government intervention in trade are usually classified as non-tariff barriers to trade and include quotas, production aids, consumer subsidies and health and safety regulations.

exported commodities, in which the country has a comparative advantage, to the imported commodities. In addition, as the prices of imports will then rise, consumption will shift away from the imported commodities to those produced domestically. The effect on the distribution of income will be to shift it from the producers of the exported good to those who produce the imported good, and away from consumers to producers.

The imposition of a tariff could, however, increase a country's welfare if the country's imports are large enough to influence world price. A tariff causes the cost of imports to rise and therefore the demand for imports to fall. The price on the world market of the imported good may therefore fall, improving the terms of trade for the importing country, as the cost of its imports fall relative to its exports. If the improvement in the terms of trade outweighs the loss in welfare due to production and consumption effects in the importing country, then the country will gain by imposing a tariff. 'World welfare' will still fall, however, as the exporting country will be worse off. However, in practice, countries have rarely applied tariffs for this reason. One notable exception, outlined in Ritson (1977), is the UK, which exploited its market power as a major importer of agricultural products for import-saving reasons during the 1960s.

Countries have, however, applied tariffs to their trade on the basis that national interest is the major consideration in policy decisions, with the implications for world welfare being of secondary importance or even a matter of indifference. Tariffs have been used for various other reasons in addition to increasing a country's welfare, for example, to redistribute income within an economy for political reasons, to protect developing domestic industries, to maintain industries for national security reasons, or as a means of raising government revenue.

THE EFFECT OF CUSTOMS UNIONS ON TRADE

The previous section illustrated that free trade under certain assumptions is the optimum market situation which maximises world welfare. In reality there is a long history of government intervention in trade, especially in the developed world, and so liberalisation of trade would be expected to increase world welfare. The effect of establishing the EC as a customs union, with no restrictions on trade between its member countries, and a common barrier to third country trade, was therefore initially thought to be a movement towards free trade which would increase world welfare. However, Viner (1950), using the concepts of 'trade creation' and 'trade diversion', illustrated that the establishment of a customs union would not necessarily increase world welfare but could actually reduce it. Viner illustrated that a customs union would increase welfare if it increased the level of trade (that is, was 'trade-creating'), and would reduce welfare if it reduced trade (that is, was 'trade-diverting'). The net effect of a customs union on trade would depend upon the production, consumption, trade and tariff structures, of the countries concerned.

Viner argued that a customs union would have a trade-diverting effect if, through joining the customs union, a country switched from a low-cost source of supply to a high-cost source of supply. For example, if two countries A and B, only one of which (B) produced a particular good, established a customs union, removing restrictions on trade between them and establishing a common external barrier to third country trade, then the importing country A would alter its source of supply of the particular good to country B. If the cost of production of the particular good in country B was greater than the previous source of supply, taking into account any tariff that country A applied previously, then the price of imports into country A would rise and the quantity of imports fall. The customs union would then have a trade-diverting effect and reduce welfare in the importing country, and in the rest of the world. Trade creation would occur if the two countries produced the same product under tariff protection and if, after the removal of the tariff between the two countries, production was concentrated in the lower-cost producer and exported to the other. The price of the commodity would fall due to the specialisation of production, the effect being greater the larger the difference in the cost of production between the two countries. The result would therefore be an increase in the welfare of the two countries and in world welfare.

Viner in his theory on the effects of customs unions concentrated on production. The theory was developed by Meade (1956), Gehzels (1956-57) and Lipsey (1960). All illustrated that the effects of a customs union would not be limited to production but would also affect demand. Thus the establishment of a customs union would cause the relative prices of products to shift in favour of commodities produced domestically or within the customs union, and therefore demand would rise for these products. The effect of a customs union on consumption therefore would be to reduce the welfare losses from a trade-diverting customs union. Lipsey illustrated that the gains from the change in consumption could outweigh the losses from the diversion of trade. A trade-diverting customs union may therefore actually increase the member country's welfare, although world welfare would be reduced.

The net effect of a customs union would be more likely to increase welfare if two countries within the union are producing similar goods yet potentially could produce complementary goods. In this situation the scope for specialisation, and therefore the gains from trade would be greater. The gains would also be greater the larger the customs union, as the scope for specialisation of production would be greater. A customs union is also more likely to result in net welfare gains if the common external tariff is lower than the average tariff of member countries prior to the establishment of the customs union. Finally, the greater the trade between member countries, and the proportion of domestically traded goods, prior to the establishment of the customs union, the greater the gains will be, as the scope for losses due to restrictions on third country trade are less, and the gains from liberalising intra-EC trade more.

THE EFFECT OF THE ESTABLISHMENT OF THE EC ON INTRA-COMMUNITY AGRICULTURAL TRADE

As the EC incorporates a customs union, it might be expected that its establishment, with the removal of intra-Community restrictions on trade and the application of a common external barrier to trade, would cause a specialisation of production within the Community. As stated previously the EC is a common market which differs from a customs union in that it aims to allow the free movement of the factors of production as well as products. The effect of a common market on trade should therefore be to encourage a greater degree of specialisation than would occur within a customs union but otherwise the effects would be the same. However, the EC has not yet achieved the free movement of the factors of production, especially in the agricultural sector. In addition the source of imports would switch from outside the Community to within it, and consumption would shift in favour of goods produced in the Community. The degree to which Community-produced commodities are substituted for those produced outside the Community would depend upon whether the common external barrier to trade was greater than the barrier to trade applying in member countries prior to joining the customs union. In the case of the agricultural sector, however, the post-EC barrier to external trade was generally higher than the pre-EC barrier; therefore the shift from extra- to intra-Community trade would be expected to be greater. Intra-Community trade would increase, possibly at the expense of extra-Community trade, depending upon whether the Community had a net trade-creating or diverting effect.

The specialisation of production within the Community would occur between agriculture and other sectors in the economy, with agricultural production being concentrated in those regions which have a cost advantage in production. In addition specialisation of production would occur within the agricultural sector itself.

In the event, the effect of the establishment of the EC on the agricultural sector was distorted by the CAP, which, by removing products from the market at established minimum prices, and by restricting extra-EC trade, maintains a relatively high level of support for the agricultural sector compared with other sectors. The effect of the CAP is therefore to inhibit the specialisation between agriculture and other sectors in the economy. In addition as the level of support varies between agricultural commodities, production will be concentrated in those commodities receiving a higher degree of support. The CAP, due to the high support prices for agricultural commodities, will also affect consumption in that a higher proportion of income will be spent on agricultural commodities than on other commodities, compared with an alternative situation of lower or no support. In addition, consumption will be expected to shift to those commodities with a relatively lower degree of support.

The effect of the establishment of the EC and the implementation of the CAP was expected to be gradual because of the transitional period over which prices and the external barriers to trade were equalised between countries. The transitional period for the original six member countries of the EC ended in 1967. However, as we saw in Chapter 3, free trade within the EC(6) continued only until the introduction of MCAs, in 1969. The expected effect of the establishment of the EC on agriculture in the Community was thus further reduced by the introduction of MCAs, encouraging production in those areas with higher relative prices and inhibiting specialisation of production.

The distortions caused by the CAP and MCAs on the specialisation of production imply that the effects of the EC membership on trade could be due more to the impact of the CAP rather than to the original objective of the EC - that of free trade between member countries. In the case of agriculture the gain from membership of the EC to any particular country would depend upon the level of support and the trading position of the country prior to the implementation of the CAP. For example, if the CAP support price was higher than that applying in a particular country prior to joining the EC, then the country would gain if it was a net exporter and lose if it was a net importer. The distribution of benefits and losses would depend upon the commodity composition of each country's trade. A country may be a net importer and yet benefit from an increase in support prices after the implementation of the CAP if it imports products with a relatively low level of support and exports those with a relatively high level of support.

The establishment of the EC would therefore be expected to lead to a specialisation of production within the member countries of the Community, both between agriculture and other sectors in the economy and between agricultural commodities in member states. The degree of specialisation would be affected by the CAP and MCAs. As a result member countries may not necessarily specialise in those commodities in which they have a cost advantage, but in those commodities receiving a relatively higher degree of support.

EMPIRICAL ANALYSIS OF THE IMPACT OF THE CAP ON INTRA-EC TRADE

There have been a number of studies which have attempted to evaluate the impact of the CAP on intra-EC trade in agricultural commodities. These fall into two categories. First, there are those which attempt to evaluate the overall impact of the CAP, that is, the removal of internal trade barriers and the establishment of a common external trade barrier. These generally try to evaluate whether the CAP has had a trade-diverting or creating effect and to assess its impact on prices within and outside the Community. The second type of analysis of the CAP attempts to measure the impact of MCAs on internal prices, production, consumption and trade.

Most studies on the general impact of the CAP have concluded that the EC has caused domestic agricultural prices to rise, production to increase, consumption to fall and world prices to fall. Therefore the level of extra-EC trade has fallen at the expense of intra-EC trade. This conclusion was reached by Dean and Collins (1967) and Zusman, Melamed and Katzir (1969) in their analysis of the market for oranges. Thorbecke (1975) in his analysis of trade in temperate foodstuffs concluded that the overall effect of the CAP was to divert total trade (that is both intra- and extra-EC trade), although trade was created within the Community (that is intra-EC trade). He estimated that trade creation occurred with products such as feedstuffs, dairy, maize and wheat, but trade diversion occurred with trade in meat, eggs, fish and barley. Saunders (1984) estimated that the trade impact of the establishment of the CAP was complete by 1967 for the original member countries and for the countries of the second enlargement by 1977. The preference for trade between the original six member countries as opposed to the nine member states after the second enlargement was highest in 1969 and remained so until 1973 when, *ceteris paribus*, a trade flow in agricultural products between the original member states was over two and a half times higher than one between the original six and the three countries of the first enlargement. Saunders also estimated, in the case of soft wheat, that in 1978/79 the CAP raised internal prices by up to 50 per cent, lowered world prices by approximately 2 per cent and reduced extra-EC trade.

Loseby and Venzi (1978) analysed the impact of MCAs on trade from two perspectives, first that of West Germany (a strong-currency country with positive MCAs) and second that of France and Italy (weak-currency countries with negative MCAs). The results of their analyses on the impact of MCAs on a country's competitiveness and comparative advantage were inconclusive but seemed to support their hypothesis that MCAs affect both the competitiveness and comparative advantage of member states. Schmitz (1979) and Saunders (1984) estimated that the elimination of MCAs would raise the EC price and lower world price and that there would therefore be a resulting loss in world welfare. The loss in welfare would be greater in those countries with negative MCAs and for commodities the production of which was concentrated in countries with negative MCAs. This result was due to the relatively high proportion of negative MCAs in the late 1970s, which implied a higher overall price level if all MCAs were removed. However, to the extent that the differences in MCAs between countries did not reflect any economic criteria for price differences, they acted arbitrarily. If these studies were repeated for more recent years the results would be expected to be different, given the different pattern of MCAs over the 1980s.

The consensus therefore reflects the theory, in that the establishment of the EC, with a higher average common price than existed previously and a common external barrier, had an overall trade-diverting impact in temperate agricultural commodities. The general lowering of common prices in real terms, and therefore the external trade barrier, over the 1980s would be expected to lower the trade-diverting impact of the CAP and increase the amount of extra-EC trade. In addition as 1992 approaches, the removal of the remaining restrictions on intra-EC trade and the mobility of factors of production would be expected to encourage the specialisation of production in agricultural commodities in the Community. and therefore to increase intra-EC trade, which has been stable or falling in recent years.

TOTAL AND AGRICULTURAL TRADE OF THE EC OVER THE POST-WAR PERIOD

The EC was established during the period when there was an unprecedented rise in world trade. Most of the increase in trade was between industrialised countries in manufactured goods, reflecting the general expansion of economic activity and the liberalisation of trade through the GATT. Agricultural trade also increased, but at a slower rate than for most other products, and its share of world trade fell from 20 per cent in 1955 to 9 per cent in 1986 (UNCTAD). This decline in the importance of world agricultural trade is possibly due to the liberalisation of trade being concentrated in industrial products, to technical innovations in industrial production and to regional specialisation. Moreover, industrialised countries have followed increasingly protectionist policies towards agriculture, restricting growth in agricultural trade.

The above trends in world trade were reflected in the EC(12), where total imports and exports expanded and the EC retained its share of world imports at 35 per cent and increased its share of world exports from 34 per cent in 1960 to 37 per cent in 1986. However, agricultural trade declined in importance, from accounting for 20 and 10 per cent of imports and exports respectively in 1960 to 11 and 9 per cent in 1986 (Eurostat 1987a).

The removal of barriers between member states did result in an even greater expansion of intra-EC trade, lending credence to the argument that the EC was trade-creating. Intra-EC trade grew from 27.6 per cent of imports and 40.8 per cent of exports in 1960 to 57.8 and 57.2 per cent respectively in 1986. Most of this increase occurred between 1960 and 1970 and among the original six members of the EC, that is, before the first enlargement, and was largely due to the expansion in the West German economy and trade.

Although agricultural trade grew at a slower rate than total trade over the post-war period, intra-EC agricultural trade increased from 31.1 and 64.6 per cent of total imports and exports in 1962 to 62.3 and 72.8 per cent in 1986. This is illustrated in Figure 9.1, which shows the trend in intra-EC agricultural trade over the period 1962 to 1986. This reflects the general expansion in intra-Community imports especially during the 1960s. Intra-EC agricultural exports also rose over the 1960s but then peaked during the 1970s and fell during the early 1980s as self-sufficiency within the Community rose and the surplus was exported to third countries.

Figure 9.1: Intra-EC(12) Agricultural Trade as a Percentage of Total Trade: 1962-1986

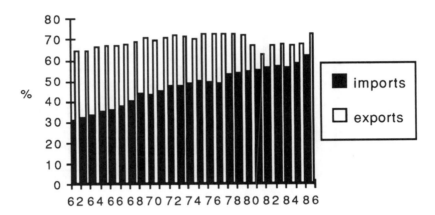

Source: Eurostat (1987a).

The Intra-EC Agricultural Trade of Individual Member Countries

The impact of the adoption of the CAP on individual member states depended upon whether they were importers or exporters of agricultural products and who were their main trading partners prior to entry.

The change in intra-EC(12) food and live animal trade by country discussed above is summarised in Figures 9.2 and 9.3, which give intra-EC(12) imports and exports as a percentage of the total. These figures show quite clearly the change in the percentage of intra-EC trade of each member state, changing in accordance to when they entered the EC. Therefore in the case of the original member states the largest increase in intra-EC trade was between 1960 and 1970, for the three countries of the first enlargement between 1970 and 1980, and for the other member states between 1980 and 1986.

France and the Netherlands, two of the original member states and agricultural exporters, experienced an expansion in exports from 1962 as they benefited from the relatively high support prices under the CAP, the common external barrier and free trade within the EC. French intra-EC exports increased substantially after entry, as illustrated in Figure 9.3, from 40 per cent of the total in 1960 to 74 per cent in 1970 although they then fluctuated to 72

173

per cent in 1986. French exports to her two main markets expanded considerably, with a doubling of exports to West Germany from 1964 to 1966 and a threefold increase of exports to Italy over the period 1968 to 1979 (Saunders 1984).

Figure 9.2: Intra-EC(12) Food and Live Animal Imports as a Percentage of Total Trade: 1960-1986

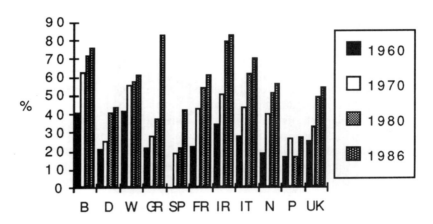

NB: No data are available for Spanish trade in 1960.
Source: Eurostat (1987a).

Figure 9.3: Intra-EC(12) Food and Live Animal Exports as a Percentage of Total Trade: 1960-1986

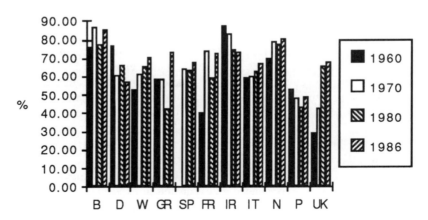

NB: No data are available for Spanish trade in 1960.
Source: Eurostat (1987a).

The intra-EC exports of the Netherlands were relatively high in 1960, accounting for 70 per cent of the total, but also rose to 80 per cent by 1986 (Eurostat 1987a) The Netherlands exports to her main market, West Germany, increased considerably from 1965 (by two and a half times to 1979), and her other EC markets also expanded but at a slower rate.

The other original member states, Germany, Belgium/Luxembourg (Belux) and Italy, were net importers of agricultural commodities when they joined the EC. However, West Germany, Belux and to a lesser extent Italy also responded to the high support prices and expanded production and exports, particularly since 1966. In addition the proportion of their exports to other member states increased from 53, 76 and 59 per cent respectively in 1960 to 71, 85 and 67 per cent in 1986. West Germany expanded production of cereals and dairy products, her main market being Italy. Belux expanded production of cereals and intensive production of meat and meat products.

Agricultural imports of the original six member states also increased but at a much slower rate than exports, reflecting the rise in self-sufficiency in these countries. The proportion of intra-EC imports increased over the whole period 1960 to 1986 but in particular between 1960 and 1970. Belux imports from the community increased from 41 per cent in 1960 to 63 per cent from 1960 to 1970 and then to 76 per cent by 1986. Likewise West Germany imports from the Community increased in importance from 42 per cent in 1960 to 61 per cent in 1986. France and the Netherlands as major exporters of temperate products, such as cereals and dairy products, had a relatively lower proportion of imports from other member states; however, even this rose from 19 and 22 per cent respectively in 1960 to 40 and 43 per cent in 1970 and 56 and 61 per cent in 1986 (Eurostat 1987a).

Denmark and Ireland were major exporters of dairy and meat and meat products when the EC was established. Denmark's main market was the UK followed by West Germany, and almost all Ireland's exports were destined for the UK market. Their reliance on the United Kingdom market meant they did not join the EC until the United Kingdom did in 1973. This may partially explain why their exports failed to increase at the same rate as the original member states during the 1960s and only increased after their entry into the EC. This trend is reflected in the percentage of Danish and Irish exports to other member states, which fell from 1960 to 1970 by 16 and 4 per cent respectively. In particular Danish exports to West Germany fell considerably from 1962 to 1973 and then rose again after 1973. After entry Danish exports to other member states also rose but never at a rate comparable to the Netherlands during the 1960s. The percentage of imports from the Community has risen from 20 per cent in 1960 to 43 per cent in 1986 with the main rise being during the 1970s after entry.

Ireland, prior to membership of the EC, had little trade with other member states (except the United Kingdom). However, since then her exports to other countries, in particular France, have increased.

The United Kingdom was the major food importer in Europe and during most of the 1960s experienced a relatively constant level of imports. However, as entry to the EC was anticipated, domestic agricultural policy altered, and support increased in the early 1970s, and as a result imports started to fall and exports increased. The entry of the United Kingdom into the EC did mean a major change in the source and destination of United Kingdom trade, particularly in the case of imports, most of which had come from Commonwealth countries. Therefore there was a large rise in the percentage of United Kingdom imports from other member states of the EC. The percentage of intra-EC imports increased slightly from 1960 to 1970 but then increased rapidly from 33 per cent of the total in 1970 to nearly half of imports in 1980 and 54 per cent in 1986. The main source of these imports was Denmark and Ireland but they have fallen in importance as a source since entry into the EC, with France and the Netherlands increasing their market share.

As UK exports of agricultural products increased, especially during the mid-1970s, the percentage of exports to member states also increased, from 26 per cent in 1960 to 67 per cent in 1986, the United Kingdom's main Community markets being France, Ireland, West Germany and the Netherlands.

The impact of the entry to the EC of Greece, Portugal and Spain was somewhat different in that much of their production did not compete directly with that of the other member states, with the possible exception of Southern Italy. These countries produce mainly Mediterranean products, which are to a large extent complementary to those produced by Northern European member states. However, these countries have expanded their production in commodities with a relatively high level of support at the expense of others. For example the self-sufficiency in barley has expanded in Spain whereas rice production has fallen (European Commission 1988). Greece, Spain and Portugal, prior to membership, were exporters of fruit and vegetables to the EC and have benefited with an expansion of intra-EC exports and they have also increased their proportion of intra-EC imports. Thus by 1980 the percentage of imports from other member states had risen, but was still low relative to the other nine member states, at 37, 22 and 17 per cent respectively in Greece, Spain and Portugal. However by 1986 the percentage of imports from other member states had increased to 83, 42 and 27 per cent respectively. The percentage of exports destined for other members of the Community also rose, especially in the case of Greece, where it increased from 42 per cent in 1980 to 73 per cent in 1986, reflecting again the fact that Greece joined earlier than the other two countries and therefore has had longer to alter her trading patterns.

There was a considerable rise in intra-EC trade between the original member states, which is not illustrated in Figures 9.2 and 9.3 as they include trade between all the current member states. Trade between the six expanded nearly three times faster than the growth in total agricultural trade between 1962 and 1970 (Saunders 1984) but slowed down during the 1970s to the average growth in total agricultural trade. This suggests that, for the EC of Six, the adjustment of agricultural trade to the CAP (that is, the effect of removing barriers to trade between the Six and the implementation of a common external barrier on trade) had mostly occurred by 1970.

The percentage of intra-EC agricultural imports in Denmark, Ireland and the United Kingdom rose over the whole period, but most of this increase was between 1970 and 1980. The fall in Danish and Irish intra-EC exports, in 1970, and the large rise in intra-EC imports in Denmark, Ireland and the United Kingdom, from 1970 to 1980, reflect the restrictions on trade between these three countries and the EC of Six in the 1960s, and the freeing of this trade after the first enlargement of the EC.

Intra-EC Agricultural Trade by Commodity

Agricultural trade by commodity for the EC(12) in 1986 is illustrated in Figure 9.4. This shows that fruit and vegetables (05) were the main import, accounting for 22 per cent of food and live animal trade. The relative importance of fruit and vegetables (05) in both intra- and extra-EC imports reflects the nature of this commodity group in that, for climatic and storage reasons, no single member of the Community can produce the diversity of fruit and vegetables (05) required for consumption throughout the year. The next most important commodity is coffee and beverages (07), which accounted for 15 per cent of total food and live animal imports in 1986 and 30 per cent of extra-EC imports, reflecting the fact that the Community has to import all its requirements of this commodity.

Other important imports are meat and meat products, cereals and dairy products, accounting for between 10 and 14 per cent of food and live animal imports. However, most of the trade in these commodity groups is intra-EC as they are temperate zone products produced within the Community.

Fruit and vegetables were also the main Community export in 1986, accounting for 19 per cent of the total. Most of this was intra-EC, reflecting the specialisation of production in the EC due to climatic factors. Meat and meat products are the next important export with most of this trade, as with fruit and vegetables, being intra-EC. Cereals are the most important commodity group in extra-EC exports accounting for over 20 per cent of total extra-EC exports in food and live animals.

The importance of intra-Community trade by commodity group is emphasised in Figure 9.5. This shows that 62 per cent of imports and 73 per cent of exports in food and live animals were within the Community. Over 80 per cent of imports in live animals, meat and meat products, cereals and miscellaneous food preparations (mainly margarine) and over 90 per cent of imports of dairy products were from other member states, reflecting the relatively high level of support for these commodities, the rise in self-sufficiency and therefore the lower level of extra-EC imports. Conversely, products which are mainly imported from outside the Community, such as coffee and beverages, fish and feedstuffs, either are not produced in the

Community or, in the case of feedstuffs, are imported as cheap substitutes for high-priced EC feedstuffs.

Overall a much higher proportion of exports of food and live animals are within the Community, particularly in those commodity groups in which the EC is not self-sufficient, for example, fish, fruit and vegetables, feedstuffs and coffee and beverages. However, in those commodities which are in surplus, in particular cereals, but also meat, dairy and miscellaneous food, extra-EC exports are more important.

The main change in the commodity composition of food and animal trade since the inception of the EC and the CAP has been in cereals, which were 21 and 13 per cent of imports and exports respectively in 1962 but fell to only 11 per cent of imports in 1986, whereas exports rose to 17 per cent in 1986, reflecting the growing self-sufficiency in this commodity.

Figure 9.4: Percentage of Total Food and Live Animal Trade by Commodity Grouping EC(12) 1986

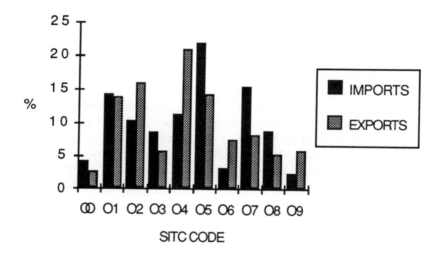

SITC Codes:

00	Food and Live Animals	05	Fruit and Vegetables
01	Meat and Meat Products	06	Sugar
02	Dairy Products	07	Coffee and Beverages
03	Fish	08	Feedstuffs
04	Cereals	09	Miscellaneous Food Products

Source: Eurostat (1987a).

Figure 9.5: Intra-EC Trade as a Percentage of Total Food and Live Animal Trade by Commodity Grouping EC(12) 1986

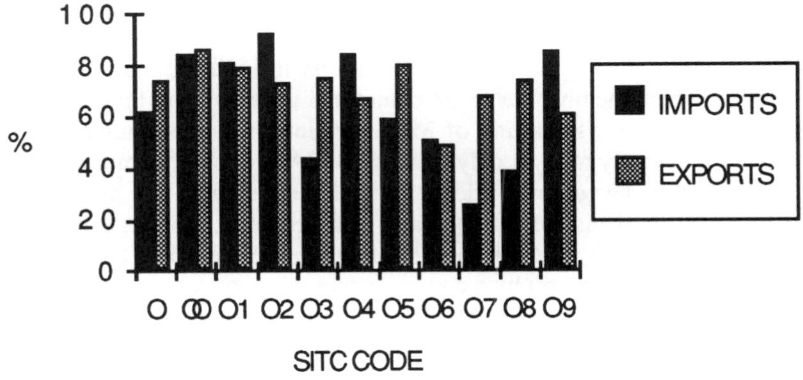

For SITC Codes, see Figure 9.4.   Source: Eurostat (1987b).

AGRICULTURAL TRADE IMPLICATIONS OF 1992

The removal of all barriers preventing the free movement of goods and services and factors of production, as envisaged by the Single European Act, would be expected to lead to an increase in specialisation of production within the Community. This would occur, not only within agriculture, but also between agriculture and other sectors, as explained earlier in this chapter. The extent to which this occurs will depend upon how far the EC achieves its objectives of a common market. Rickard (1990) states that it is not yet known what form the single market will take and the possibilities range from the EC 'maximalist' approach to the British Government's 'minimalist' approach. If the latter prevails then 1992, he argues, will be a 'non-event'.

In theory the existence of the CAP should have led to fewer restrictions on agricultural trade than in other sectors, as claimed by the Commission (European Commission 1987). The very existence of a common policy for agriculture with no barriers to internal trade, a common external barrier and common support policies implies that specialisation is more developed than in other sectors of the economy. Therefore the movement towards a common market should have less impact upon agriculture than on other sectors. However, in practice, the common policy has not prevented a wide range of restrictions, both physical and technical, applying to intra-EC agricultural trade. This is exaggerated when trade in food and drink is included, [4] with Caspari (1990) citing a study commissioned by the Community which

───────────────

4   The impact on trade in food products is considered in Chapter 11.

179

identified over 200 non-tariff barriers applying to trade in food and drink. Thus, removal of restrictions in agricultural trade is not as straightforward as the Commission claims, with *Agra-Europe* going so far as to suggest that the agricultural sector is the most restrictive and may prevent the completion of the programme towards a common market in 1992 (*Agra-Europe* 1989).

The most documented of physical barriers to intra-EC agricultural trade are MCAs which member states are committed to remove by the end of 1992. The impact of the dismantling of MCAs on intra-EC agricultural trade was discussed previously and, as only negative MCAs are now permitted, it might be expected that the common price level will rise, although this most probably will be offset by decisions at the annual price reviews. In fact a mechanism already exists whereby the first quarter of any impact on prices of the removal of a negative MCA is offset by a corresponding fall in prices.

The unknown factor when MCAs are dismantled is the mechanism which will exist when the market rate of exchange of a member state alters. At present, in the case of an alignment of EMS rates, the resulting MCAs, are dismantled over three years. This system could continue. However, it is further complicated by the 'central rate correcting factor' or the 'switchover mechanism' introduced in 1984. This mechanism effectively eliminates positive MCAs and creates negative MCAs when there is an alignment of the EMS. It was introduced to protect West German farmers from falling support prices, as the mark appreciated relative to other member states currencies, and it is therefore doubtful whether its removal would be politically acceptable.

In the case of countries which are not full members of the EMS and therefore have fluctuating exchange rates, the removal of MCAs will cause institutional support prices to alter as the currency fluctuates. This not only leads to greater uncertainty within the agricultural sector and considerable administrative problems, but also to the speculative movement of goods across borders. Therefore the entire removal of the MCA system seems unlikely, although, if it does remain, considerable practical problems will exist if customs posts at frontiers are removed.

The removal of physical barriers to trade (customs posts) has important implications also for the removal of technical barriers to trade. At present there are a number of technical barriers, including health and safety regulations and common standards, which involve physical controls at border frontiers and documentation for trade. Therefore, if border custom posts are abolished, these differences would be ineffective.

How far the Community will achieve the removal of physical barriers by 1992 is questionable. *Agra-Europe* (1989) emphasises that the Commission White Paper does not outline how this is to be achieved.

The Commission aims to overcome technical barriers to trade in the form of health and safety regulations and standards either by achieving a common European regulation and/or standard, or, where different standards exist, between member states, by making these acceptable throughout the

Community. This is an area which will be fraught with controversy and the Single European Act did not overrule Article 36 of the Treaty of Rome, which permits member states to restrict trade on health grounds. Even if the Community can achieve the harmonisation of regulations there are problems in ensuring that these are applied uniformly throughout the Community. In addition there are important implications for third country trade where, if the Community is considered as one country, an outbreak of disease in one member state may cause an import ban for all Community countries.

CONCLUSION

The descriptive analysis of EC trade over the post-war period substantiates the following conclusions. The changes in agricultural trade reflect two main factors influencing the trading patterns. First, the changes in policy with, for example, a concentration of production in those commodities attracting relatively higher levels of support. Second, trade was influenced by the technological advances in agriculture over this period enabling expansion of production and therefore a rise in self-sufficiency. This led to surpluses in certain commodities and problems in their disposal.

Agricultural trade in the EC has grown considerably since World War II though more slowly than total trade. Intra-EC trade in agricultural products expanded faster than the rise in total agricultural trade especially during the 1960s. Most of this increase was in imports whereas, although intra-EC exports rose over most of the period, they actually fell, as a percentage of total agricultural trade, during the late 1970s and early 1980s. This fall was due to the rise in self-sufficiency in agricultural products, and the rise in world commodity prices over this period, encouraging extra-EC exports.

Intra-EC trade also reflected the changes in policy as countries joined the Community. Thus, for the original six member states, intra-EC trade grew at the fastest rate during the 1960s; for the three countries of the first enlargement during the 1970s; for Greece in the second enlargement during the late 1970s and early 1980s; and for the two countries of the third enlargement during the 1980s. In the case of countries which did not join at the outset, their intra-EC trade, particularly in the case of exports, frequently fell prior to joining the Community.

The commodity composition of trade also reflects the above factors, with a rise in the exports of products produced and supported in the EC. The most important imports are now those which the Community cannot produce, or are not available domestically throughout the year due to climatic factors. Other important imports are in commodities which do not receive a high degree of support, such as fish and fodder. Intra-EC imports are typically more important in those commodities which are in surplus, as the need for extra-EC imports falls, and less important in those commodities for which the EC is not self-sufficient. This is also reflected in exports, with extra-EC exports being important in those commodities for which there is a surplus and less so for those which there is not.

The future developments in intra-EC trade will obviously depend upon the way the CAP develops and the effects of 1992. It is too early to estimate the impact of policy changes during the 1980s, such as milk quotas and stabilisers, but it can be expected that these have had a limited impact on intra-EC-trade as they were primarily designed to reduce extra-EC exports. However, there are a number of future developments which would be expected to encourage the specialisation of production. These include the removal of MCAs, the lowering of intervention prices and introduction of stricter regulations for intervention buying, and the freeing of movement in factors of production.

REFERENCES

Agra-Europe (1989) *1992: Implications for the Agri Food Industry.* Special Report No. 42, *Agra-Europe.*

Caspari, C. (1990) *The 1992 Process in Food and Agriculture: the European View.* Paper presented at the Agricultural Economics Society Conference, Wye College, University of London.

Dean, G.W. and Collins, N.R. (1967) *Trade in Fresh Oranges: an Analysis of the Effect of the EEC Tariff Policies.* Giannini Foundation Monograph, University of California, Berkeley.

European Commission (1985) *Completing the Internal Market,* White Paper, Luxembourg.

European Commission (1987) *Europe Without Frontiers Completing the Internal Market,* European Documentation 4/1987.

European Commission (1988) *Agricultural Situation in the Community,* Luxembourg.

Eurostat (1987a) *External Trade,* Statistical Yearbook, Luxembourg.

Eurostat (1987b) *NIMEXE,* Luxembourg.

Gehzels, F. (1956-57) Customs Unions from a Single Country Viewpoint, *Review of Economic Studies,* 24(1).

Lipsey, R. (1960) The Theory of Customs Unions: a General Survey, *Economic Journal,* 70(3), pp 496-513.

Loseby, M. and Venzi, L. (1978) The Effect of MCAs on EC Trade in Agricultural Commodities, *European Review of Agricultural Economics,* 5(3), pp 361-379.

Meade, J.E. (1956) *The Theory of Customs Unions,* North Holland Pub. Co., Amsterdam.

Rickard, S. (1990) *The Competitiveness of UK Agriculture in the Single Market,* Paper presented at Agricultural Economics Society Conference, Wye College.

Ritson, C. (1977) *Agricultural Economics: Principles and Policy,* Blackwell, London.

Saunders, C.M. (1984) *Intra-EC agricultural trade with special reference to wheat.* Unpublished PhD thesis, Department of Agricultural Economics and Food Marketing, University of Newcastle upon Tyne.

Schmitz, P.M. (1979) EC Price Harmonisation: a Macroeconomic Approach, *European Review of Agricultural Economics,* Vol. 6 No. 3, pp 165-190.

Thorbecke, E. (1975) The Effects of Economic Integration on Agriculture. In: Balassa, B. (ed.) *European Economic Integration,* North Holland Pub. Co., Amsterdam.

Truman, M.E. and Reswick, S.A. (1974) The distribution of western European trade under alternative tariff policies, *Review of Economics and Statistics,* 56, pp 83-91.

UNCTAD *International Trade Statistical Yearbook,* Vols 1 and 2. Various issues.

Viner, J. (1950) *The Customs Union Issue,* Carnegie Endowment for International Peace, New York.

Zusman, P., Melamed, A. and Katzir, A. (1969) *Possible Effects of EEC Tariff and Reference Policy on the European and Mediterranean Markets for Fresh Oranges,* Giannini Foundation Monograph No. 24, University of California, Berkeley.

CHAPTER 10

THE CAP AND RESEARCH AND DEVELOPMENT POLICY

David Harvey

INTRODUCTION

'Whoever could make two ears of corn or two blades of grass to grow upon a spot of ground where only one grew before, would deserve better of mankind, and do more essential service to his country than the whole race of politicians put together.' (Jonathan Swift - *Gulliver's Travels,* Voyage to Brobdingnag)

As repeatedly referred to in this book, the Objectives of the Common Agricultural Policy lay the foundation for the EC agricultural policy. These objectives:

'1. to increase agricultural productivity by promoting technical progress and by ensuring the rational development of agricultural production and the optimum utilization of the factors of production, in particular labour;

2. thus to ensure a fair standard of living for the agricultural community, in particular by increasing the individual earnings of persons engaged in agriculture;

3. to stabilize markets;

4. to assure the availability of supplies;

5. to ensure that supplies reach consumers at reasonable prices '

(Article 39, Treaty of Rome)

make clear that promotion of technical progress and improvement of agricultural productivity are fundamental to the development of European agriculture. It would seem to follow that a common approach to agricultural research and development would form a key part of the collection of policy instruments which support these objectives. However, this is not the case. Agricultural research and development (R&D) remain very much a national responsibility, along with a substantial amount of agriculturally related support and regulation, especially concerned with inputs, capital investment, taxation and structural change. Indeed, total taxpayer spending on agriculture is almost equally divided between the European budget and national exchequers, a fact which is frequently ignored in discussion of European agricultural policy. The common features of the CAP are almost solely restricted to intervention in product markets.

In fact, the present climate of surplus production and concern over the fate of the countryside seems to suggest that R&D is the enemy of both the CAP and the countryside. This is a remarkable thought, contrasting dramatically with the Swift quote above and almost obscenely with the paradox of continued malnutrition and chronic famine in many parts of the developing world. Yet the development of European agriculture as an increasingly intensive and industrial activity is leading both to a growing gap between the amount that is produced and the amount that European consumers are prepared to consume, and to serious concern about the fate of natural habitats, traditional landscapes, clean rivers and species survival. This development is in turn heavily supported and encouraged by so-called technological improvements. Furthermore, much of the more obvious technological change has occurred through improved machinery, plant and equipment which are only economically viable if labour forces are reduced and farms and fields get bigger. The small 'family' farm and its larger neighbour with large labour forces, both supporting a traditional village and rural economy and society, are becoming historical artefacts, to the apparently obvious detriment of rural life and the attractiveness of the countryside as a place to live and play. Yet this contrast between technological change, which simultaneously increases unsaleable surpluses and resulting taxpayer costs, and also damages the natural and social rural environment, is only part of the reason for concern about agricultural R&D.

Bovine Somatotrophine (BST) is the latest in a long line of impressive technological changes, but provides a classic illustration of another dilemma. BST is a hormone which is produced through genetic engineering. When incorporated in dairy cattle feed, it has the effect of increasing the dry matter intake of the cow and thus, under favourable conditions, increasing the milk yield of the cow. The production of the hormone from 'artificial' engineered microbes, despite that fact that it is indistinguishable from the same hormone produced naturally within the dairy cow itself, causes considerable concern amongst the general population. May there not be serious side-effects from the administration of this 'drug' which is bound to get into the milk we drink?

It is hard to think of a farm product with a more complete image as a natural food than milk, and this technology threatens that fundamental concept as surely as the *Salmonella* 'crisis' did for eggs in the United Kingdom in 1988. But exactly the same effect could, at least in principle, have been achieved through the conventional breeding of dairy cows so as to increase the amount of BST produced naturally within the cow itself. Would this cause as much consumer resistance, even if they were told of the reasons for dairy yield improvements? In addition, of course, the possible introduction of BST also illustrates the apparent madness of advancing technology. What is the point of trying to produce even more milk when we cannot use what is already produced? Quota restrictions on milk production coupled with dramatically increased dairy cow yields [1] can only lead to fewer dairy cows and fewer dairy farms. What will this do to the production of beef, much of which comes from the dairy herd, and what will it do to the 'traditional' small dairy farm?

A final example illustrates the effects of technological change directly on policy. The European Community banned growth promoters in meats, especially beef, in 1987. These hormones were used to increase the growth rates of beef animals. Typically they are based on testosterone (the male hormone), produced naturally in the entire male animals, so that the ban might be expected to encourage the use of entire males for 'bull beef' production. It is difficult to determine the origin of these hormones in carcass meat, so that a ban on artificial hormone use is difficult to police. The scientific evidence about the possible harmful effects of these hormones is also disputed. Nevertheless, the EC (Parliament, Commission and Council of Ministers) were concerned enough about the possible harmful side-effects of hormones in meats to impose the domestic ban, responding to genuine consumer worries. This ban was to be extended to imports of meat at the beginning of 1988. However, following complaints by traditional exporters to the EC (especially the United States), the imposition of this ban was delayed until January 1989 to allow negotiations about possible compromises or compensation for loss of trade. The US claimed a) that there was insufficient scientific evidence of harmful side-effects to warrant the ban; b) that the ban was not being policed within the EC so that the ban on imports was simply a trade restriction. Negotiations within the GATT are still going on over this issue, and have become entangled in the more general negotiations over trade distortion and agricultural protection under the Uruguay round of GATT talks (see Chapters 12 and 14). The major points to be drawn from this example are that technological change within agriculture is increasingly a matter of concern to consumers as well as to farmers, and also that responses to technological change affect farm and agricultural trade policies directly.

In view of the complications caused by, and policy interactions associated with, technological change and the underlying R&D, there are serious questions about the possible benefits of R&D, and what sort of policy ought to be followed in funding and controlling agricultural R&D. This is a huge area, which can only be partially covered in this chapter. The purpose of the rest of the chapter is to explain the underlying economics of R&D as it interacts with agricultural support policy under the CAP, and to offer some concluding comments on the role of R&D. The core of this chapter deals with three major areas:

The 'Modelling' of Technological Change - how economists deal with technological change and the implications of this treatment;

---

1   Yield increases of as much as 25 per cent have been suggested following introduction of BST.

The Consequences of Technological Change - which depend on market and policy conditions surrounding new innovations;

The CAP and Technological Change - the past history of the CAP and its interaction with R&D and the implications of the CAP policies for technological change and for the underlying R&D effort.

THE 'MODELLING' OF TECHNOLOGICAL CHANGE

In order to deal with the effects of technological change, some underlying characterisation of the supply conditions within agriculture is necessary. Economists start with the notion of the supply curve (Figure 10.1).

Figure 10.1: The Supply of Agricultural Products

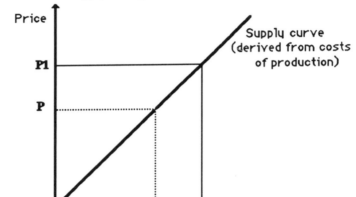

This pictures the relationship between the price of a commodity and the quantity which the industry may be expected to supply, *holding all other factors and influences on farm supply quantities unchanged.* It is generally regarded on both theoretical and empirical grounds as being upward sloping, that is, the greater the quantity which is required from the industry, the higher the price will have to be, or, conversely, the higher the price of products, the greater will be the quantity supplied to the market. Although some argue that farmers will be content to supply the same quantity or even less as prices are increased (or produce more as prices fall to preserve income levels) there is virtually no sound empirical evidence to support this argument. Furthermore, logic does not support this contention. Higher prices justify greater expenditure on production practices and inputs with a resulting increase in the quantities supplied. To get more output from the industry, farmers have to work harder, farm land more intensively, apply more fertilisers and chemicals, use more machinery and labour, and use less suitable land. All this costs money, and farmers will not undertake this extra effort and spending without the necessary additional return. Conversely, if product prices fall, then some farm expenditures and investment can no longer be justified by the returns and will be discontinued, thus reducing output levels. In order to persuade farmers to produce Q2 rather than Q1, we will have to pay them P2 rather than P1. Conversely, if prices are set at P2 then we would expect them to supply Q2, while if price is P1 then quantity supplied (other things being equal) will be Q1.

What does technological change (the product of the R&D system) do to this picture? The first point to notice is that R&D only affects supply as and when the new technology is adopted by at least some in the industry. For this to happen, the new technique (or improved input) must offer some commercial advantage over existing practices and processes. As and when new technology is adopted, it effectively *shifts* the supply curve, as in Figure 10.2.

Figure 10.2: The Effects of Technological Change on the Supply of Agricultural Products (a)

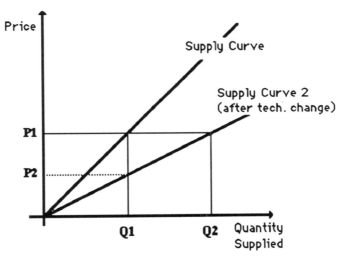

Technological change effectively allows farmers to reduce the costs of production; if it does not then it will not be adopted. Thus, for each and every quantity, the costs of production are reduced by technological change, which shifts the supply curve downwards as shown in Figure 10.2. This allows either more to be produced at the same price (Q2 rather than Q1 at price P1) or allows the same quantity to be produced at a lower price (Q1 at price P2 rather than P1).

The implication of this representation is that every adopted technological change offers the opportunity for some combination of cost and price reduction combined with some output increase. There is no obvious distinction between 'cost-reducing' and 'output-increasing' technological change. However, some technological changes may shift the underlying supply curve in different ways from others, so that the effects may be different, while the conditions in the market-place, as influenced by policy, will dictate the consequences of technological change. [2] Thus the elementary economics of technological change are illustrated in Figure 10.3.

Figure 10.3: The Effects of Technological Change on the Supply of Agricultural Products (b)

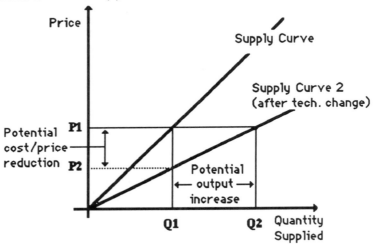

One final point to notice about this representation of technological change is that the shift in supply does not occur overnight. The process of adoption of new technologies is likely to be important and leads to shifts in supply which happen more or less continuously over time. The 'early adopters' benefit from the reduced costs and increased profits offered by the new technology before the consequences of the ultimate increase in supply filter through from the market. The 'laggards', however, may be faced with the market consequences before they have adopted the new techniques and thus before they have adjusted their costs

2   Although technological change is pictured here as a pivotal shift of the supply curve around its intercept with the axes (origin in this special case), there is no reason to suppose that this is the only sort of technological shift which is possible. Some other sorts of shift are explored by Lindner and Jarrett (1978). However, for the purposes of this exposition, the rotational shift serves as an unrestrictive illustration of the effects of R&D on agricultural supply. Some additional implications of this approach and an elaboration of alternative ways of representing technological change are provided by Wise (1984) in a criticism of the analysis provided by Lindner and Jarrett (1978).

of production to become competitive and viable under the changed market conditions. These farmers may well find themselves trapped on the 'technological treadmill', [3] on which farmers are 'forced' to adopt new technologies merely to maintain their incomes and remain viable rather than encouraged to adopt to improve their incomes and profits. The industry is obliged to run faster to stay in the same place through the consequences of technological change. What are these consequences?

THE CONSEQUENCES OF TECHNOLOGICAL CHANGE

If we consider first a country market isolated from any trade (with no imports or exports) then market prices are determined by the interplay of domestic supply and demand. Figure 10.4 represents this situation, with the quantities demanded increasing slightly as farm product prices are reduced (other things being equal). In the technical language of economics, the demand for farm products is 'inelastic', with quantities demanded being rather unresponsive to price changes. In these circumstances, technical change leads to substantial price reductions and only limited increases in output and consumption. In Figure 10.4, quantity Q1 is produced and consumed prior to the technological change while the new innovation shifts the supply curve downwards to give the new market equilibrium at the lower price (P2) and the larger quantity produced and consumed (Q2). Because the price reduction is proportionately greater than the corresponding increase in output quantities, total producer revenue (and consumer expenditure), measured as price times quantity, falls as a result of technological change.

Consumers and users clearly gain substantially from technical change under these circumstances, hence Christopher Ritson's argument that objective 1 of Article 39 of the Treaty of Rome benefits consumers to a larger extent than producers (Chapter 7). Farmers in this sort of market-place find that the advantages of technical change are eroded by market changes, and become captured by the 'technological treadmill'. One important implication of this analysis is that the funding of the underlying R&D should be financed by the consumers and users rather than the producers, which usually means funding through taxpayers (that is, publicly funded R&D) since it is not otherwise practical or effective to extract payment from such a wide and disperse group. This has been the traditional argument in favour of public funding of the agricultural R&D effort.

As noted above, some farmers are more innovative than others for a variety of reasons (which are not well understood). Early adopters reap cost advantages before product prices respond and capture considerable benefits, albeit on a transitory basis. Late adopters, on the other hand, find themselves forced to adopt new technology or face bankruptcy, and hardly benefit from the

---

3   A term coined by Cochrane (1958).

technology at all. As a consequence, periods of rapid technological change can be associated with rapid structural change in the industry as early adopters grow at the expense of laggards.

Figure 10.4: Consequences of Technical Change in a Closed Market

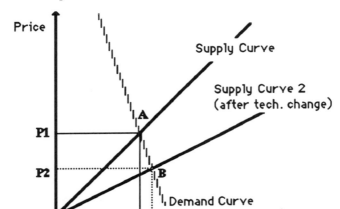

The consequences of technical change for a small *trading* country are rather different. The definition of 'small' in this context simply refers to the proportion of total world production and consumption which is accounted for by this country. So long as these proportions are small, then changes in domestic production and consumption levels have no significant effect on world market prices. Under these conditions, we may treat the world price as fixed, at P in Figure 10.5.

Technical change then results in an unambiguous expansion of domestic production (from Q to Q2) with no change in domestic market prices and no effect on consumers, aside from perceptions of the acceptability of the techniques. There is either a reduction in imports or an increase in exports of the product in question and a clear gain to the country's balance of payments. Farmers benefit from such an improvement, and might be expected to contribute to the funding of R&D, through either the private or the public sector, as they already do in France through a land tax.

However, there are two important exceptions to this case. First, some technical changes are equally applicable abroad and at home. If the technical change which causes the domestic supply curve to shift downwards is repeated in the rest of the world, then world prices will fall as a result (since the analysis contained in Figure 10.4 applies by definition at the world level, which must be a closed market). Second, the same effect (world prices falling) occurs if the country is large enough to influence world prices (as is the case for the European Community as a whole). If a large country imports less or exports more as a result of technical change, then the world market has to sell less or buy more, which will only happen if (world) prices fall. Figure 10.6 illustrates the consequences, with world prices falling from P to P2 as a result of the technical change and supply shift. In this case, consumers and users do

Figure 10.5: Consequences of Technical Change in a Small Open Market

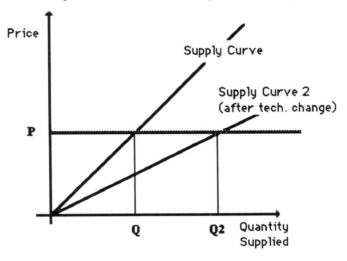

Figure 10.6: Consequences of Technical Change for a Large Trading Country or with International Technology Transfer

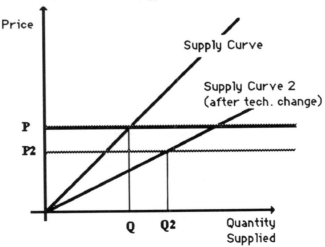

gain from R&D, while the effects on domestic producers depend critically on the effects of domestic R&D on world prices, either through the transfer of technology or through the effects of domestic supplies on world markets.

The consequences of technical change clearly depend on market conditions as well as on the nature of the new technology. However, in agriculture these market conditions are heavily influenced by farm policy, particularly the CAP.

THE CAP AND TECHNOLOGICAL CHANGE

The same analysis can be used to demonstrate both the history of the CAP and the policy response to date. Figure 10.7 illustrates the effects of technical change on a market with fixed domestic prices. It looks the same as Figure 10.5 (the small trading country case) but with the important distinction that the price (P) is now **not** the world trading price but is the domestic support price, which as demonstrated in Buckwell's chapter on the CAP and international trade, is substantially above the world trading price.

Figure 10.7: Technical Change and the CAP - History

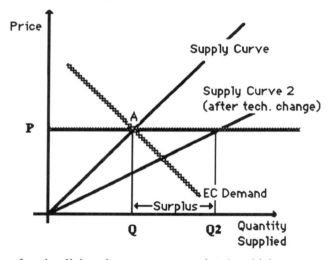

Suppose, for simplicity, that we start at point A, which represents exact self-sufficiency within the European Community; that is a position of no exports and no imports. Now introduce technical change, which shifts the supply curve to supply curve 2. If support prices are not changed, then agriculture now produces Q2 rather than Q, but demand does not change, so only Q is used, and the difference accumulates as surplus to domestic requirements. European intervention stores fill up with the produce of domestic technical change, and the costs of surplus disposal grow, not only because of the increasing surpluses but also because the EC's dumping of these surpluses on world markets depresses world prices, so the export subsidies have to be increased as well.

There are three possible responses to this situation which neatly encapsulate the present problems of the CAP. First, support prices can be reduced to re-establish domestic market balance (Figure 10.8), in which case domestic consumers gain from the R&D, though there are problems for adjustment in the European agricultural sector, and there may still be a strong case for opening up the European market to world trade. In fact, support prices within the EC have been reduced, at least in real terms and over the last few years, but not by enough to achieve domestic market balance, still less to approach world market prices, apart from in exceptional years. It is the damage which price reductions do to farm incomes, especially of those who have not adopted the latest technology, which has so far prevented the EC from following this policy option to the necessary extent.

Figure 10.8: Technical Change and the CAP - Support Price Reduction Option

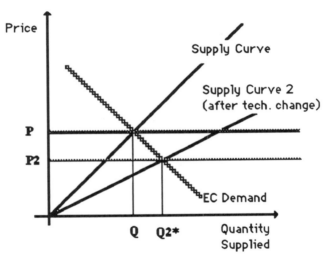

The second policy response is to restrict supplies, using quotas (Figure 10.9). This response has already been adopted in the dairy sector with reasonable success at least in so far as the budgetary costs are concerned. It is possible to interpret the cereal set-aside programme as a first step towards supply control in the cereals sector, though the European Commission and others do not accept this interpretation. In the case of the supply control option, the benefits of technical change accrue to the owners of the quotas (the licences to produce) rather than the consumers and users, in contrast to the price reduction option. The shaded area in Figure 10.9 represents the cost advantages accruing to the producers of the quota-controlled output, which results in increased demand for the licence to produce (the quota) and an increase in the market value of the quota to reflect the technologically induced cost reduction.

The third policy option is to stop technical change or, more practically, to stop the public funding of agricultural R&D.

Figure 10.9: Technical Change and the CAP - Supply Control Option

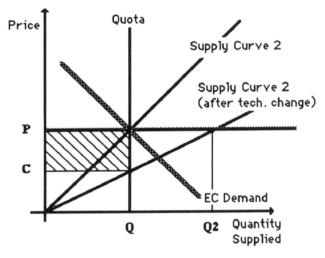

R&D EXPENDITURE ON AGRICULTURE IN THE UK

As was pointed out at the beginning of this chapter, R&D spending is not a European matter at present, it is a national responsibility. The recent UK government cuts to R&D and their current proposals can be interpreted as a response to the current problems of the CAP. Figure 10.10 charts the development of publicly funded R&D in real terms (having removed the effects of general inflation) and also illustrates the corresponding changes in the real volume of agricultural output and net product (the value of output less the value of inputs, including capital inputs, required to produce this output) again measured in volume terms, which involves holding the prices of outputs and inputs constant at some base period level.

This interpretation is, of course, over-simplistic, though as far as the British Treasury is concerned no doubt contains elements of the truth. There is little point in paying out on the one hand for the disposal of surpluses and on the other for more R&D which apparently leads to their further accumulation. Most of the cuts to public funding of the Agricultural and Food Research Council (AFRC) have been to the applied end of research. The intention has been that those parts of the associated research which the industry finds immediately useful should be funded by industry rather than from the public purse. This is a perfectly justifiable course of action, on the basis of the analysis presented here, *providing* that the policy option of reducing support prices is assumed to be unacceptable.

If, however, support prices are reduced in response to technical change in the industry, then it is the consumers who will gain, and in general it will be difficult for the private sector to recoup their costs from consumers. The upshot is that less R&D will be performed and the consumer will lose out. In addition, to the extent that the UK is the only country within Europe which follows this option, UK agriculture may find itself at a disadvantage in the future compared with competitors in Europe. Furthermore, since technical change offers a realistic way of reducing support prices without undue or unfair hardship to farmers, continued R&D provides a means of achieving greater parity with world prices and competitive ability with suppliers in the rest of the world. There is little doubt that the technological understanding which has been developed in Europe over the recent past allows European agriculture to compete more effectively with the rest of the world, even if market support policies have not allowed this competitive ability to be demonstrated or exploited.

Figure 10.10: R&D Spending and Agricultural Output Growth

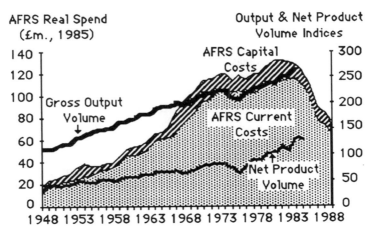

Source: Harvey (1988).

However, if supply control (rather than price cuts) is to form the major policy development within the Community, then there is a great deal of logic to the argument that producers should fund a major part of 'near market' R&D, either directly or through the input supply chain. Furthermore, since the gains to producers accrue in proportion to the ownership of the licences to produce (the farm quotas), a tax on the value of these quotas represents a fair and equitable way of raising funds for R&D. An extension of this argument would relate public contributions to 'near market' R&D to support price reductions - the greater the reductions in support prices, the greater the public contribution to R&D.

This suggestion, however, attributes more logic to the public decision-making process than is yet evident from the behaviour or utterances of the policy-makers, as explored in Chapters 3 and 6. If the Community is serious about a common market, which the existence of the CAP and current preparations for 1992 suggest, then it surely makes sense to organise public funding of R&D on a Community rather than national basis. To this author's knowledge, no suggestion for this has yet been made, in spite of the fact that virtually all private sector R&D is only justifiable on a multinational basis. However, the AFRC are beginning to expand their European links (AFRC Annual Report, 1987/88, p 5), which is a tentative step in the right direction.

COMPLICATIONS AND INTERACTIONS WITH THE CAP

The analysis presented so far has been extremely simplistic. In particular, with reference to the examples given in the introduction, no distinction has been made between the output under the old technology and that under the new. In the language of economics, the outputs or products are treated as being 'homogeneous'. Consumer concerns about the incorporation of 'artificial' hormones and genetically engineered chemicals in foodstuffs indicate that this assumption is unrealistic, at least as far as consumers' perceptions are concerned. It is not easy to incorporate this distinction in the diagrammatic analysis. However, such concerns clearly exacerbate the conflict between producers and consumers arising from technological change. As we have seen, even without this complication, technological change tends to benefit either consumers or producers, but not generally both together. When the 'new' product (produced through the new, lower-cost technology) is regarded by the consumer as being an inferior or even dangerous product, in spite of the fact that it is practically indistinguishable, the potential for conflict is even greater. Since governments will typically be required to resolve these conflicts through policy intervention, technological change and the R&D that goes with it become an urgent policy consideration.

As soon as new technology is placed on the political agenda it becomes entangled with the other objectives and instruments of policy, as has been the case with growth promoters and the CAP. In this case, it can be argued that part of the reason for the EC ban is to restrict the growth in production and thus to maintain farm prices (although not farm incomes) and that extending the ban to imports merely reinforces the domestic market support. It is difficult to tell whether this objective constitutes a major or an insignificant reason for the ban. The potential for policy conflict and argument is, however, greatly increased.

A second complication is the effect of policy and market conditions on the type and rate of technological change. There is little doubt that at least the adoption of new technologies, if not the production of these new techniques by the R&D system, is driven by market and policy signals. [4] It is hardly surprising that relatively high market prices for cereals under the CAP, coupled with substantial capital grants for 'land improvement' (drainage, hedgerow removal and so on) and significant tax advantages for agricultural capital investment, often provided by national governments, have led to increased intensification of cereal production. This has involved technological changes concentrating on improved machinery and increased use of chemicals and fertilisers to take advantage of the potential of new cereal varieties. It is an interesting question as to whether different market and policy conditions would have generated the same sort of new technologies or the same sort of adoption responses. It is far from clear that the recent technological change in the agricultural sector was inevitable and merely a consequence of the arbitrary, and historically well-funded, discoveries of scientists.

This leads to a distinction between the **social** and **private valuations** of a change. The private (sum of individual) valuations of technological change will not generally equal the social valuation. Farmers collectively might be expected to be prepared to pay more for technological change which improved wheat production and profitability than the country would pay, because the farmers do not (yet) bear the full costs of surplus disposal under the CAP, or of the possible environmental damage. The full elaboration of the consequences of further improvements in wheat yields would include the effects on the fertiliser, chemical and machinery industries, and also on asset values and labour associated with cereal production.

The social valuation of these effects is beset by two major problems: first, the appropriate weighting of the different groups who stand to benefit or lose from the improvement; and second, the status of existing policy intervention in the market. Since farmers are actually required to pay for the technological change, through royalties on the improved seed, then some of their potential benefit from the improvement will be captured by the plant breeders and their shareholders. The distribution of the benefits of R&D may be expected to affect the social valuation of the technological change, while the appropriate shadow prices to employ in the valuation of each benefit/cost component depend crucially on the view taken of the permanence or otherwise of the current policy intervention, in this case the CAP.

Further examination of the gains and losses consequent on the introduction of new technology leads to considerations of the way in which the supply curves for products shift, or equivalently of the way in which cost curves are altered (Lindner and Jarrett 1978, Wise 1984), and also of the consequences in

---

4   This is the theory of 'induced innovation'.

the input and factor markets, as explored, for example, by Schmitz and Seckler (1970). Such considerations prompt prescriptions for a more systematic and system-wide approach which seeks to adjust institutional constraints and the economic environment to favour the uptake of desirable technologies and to discourage potentially damaging technologies (see, for example, McInerney 1984). However, to date no such systematic integration of R&D effort or policy with rural or agricultural policy has developed within the European Community.

THE EUROPEAN COMMUNITY AND AGRICULTURAL R&D

While the point has already been made that most agricultural R&D in Europe is organised at the national rather than the European level, the European Community does devote effort and resources to R&D, including agricultural and food R&D. This Community-wide research occurs through two conceptually and practically different routes. On one hand, a (usually small) part of national research budgets is devoted to multilateral research projects involving research staff and resources in several different countries. Ireland, as might be expected, devoted almost 40 per cent of its agricultural research budget to such multilateral programmes in 1983 (Eurostat 1985, pp 197-198), though the equivalent proportions for other countries was much smaller, being only 3 per cent for the UK. On the other hand, the EC, through the European Commission, also commissions research within the Community, paying all (direct action projects) or part (indirect action projects) of the costs. In 1983, 'Commission' research amounted to 2.2 per cent of total civil R&D spending throughout the Community (Eurostat 1985). In addition to representing a marginal contribution to the total research effort, the Commission's involvement in agricultural and natural resource R&D is complicated by the fact that it is administered under DG VI (Agriculture), DG XI (Natural Resources) and DG XII (Science, Research and Development). Nevertheless, the programmes encouraged under Commission auspices do serve to develop co-operation and common understanding of research issues throughout the Community.

The two major programmes affecting agriculture are: a) the Standing Committee on Agricultural Research - SCAR - programme, administered through DG VI, b) the Forecasting and Assessment in Science and Technology - FAST - programme, administered by DG XII. SCAR, consisting of the directors of national research services in all EC countries, was established in 1974, to co-ordinate research throughout the Community and to initiate co-operative research. It operates through a series of five-year programmes. The most recently completed programme (1984-88) is reported in European Commission (1988). The current (fourth) programme runs from 1989 to 1993 and its aims, priorities and organisation are reported in European Commission (1990).

SCAR is explicitly responsive to problems and issues raised through the CAP. In the words of the European Commission (1988):

'The Commission's green paper entitled "Perspectives for the CAP" issued in 1985 and the subsequent memorandum, *A future for Community agriculture*, identified the deficiencies of the the CAP, particularly with regard to the creation of agricultural surpluses, and indicated its intention to promote research and extension designed to assist the diversification and re-orientation of production in line with market demand. Accordingly, in April 1986, the Commission submitted a new proposal to the Council pinpointing specific areas of research which could help solve the surplus problem. At the same time it drew attention to the contribution that research could make in developing viable farm practices which would also be acceptable in environmental terms. Subsequently, these ideas have been developed further in the plans put forward of the next five year period, 1989-93.'

The five-year programmes consist of support for research co-ordination (through seminars and workshops of scientists and scientific exchanges between countries) and for common research projects aimed at specific problems and conducted by groups of researchers from several member countries. The specific problems identified for the third programme (1984-88) were: utilisation and conservation of natural resources in agriculture, particularly energy, land and water; structural problems - Mediterranean, other less-favoured areas, agro-food issues; improvement of animal and plant productivity, including animal health and welfare. The fourth programme is directed towards a) conversion, diversification, extensification and protection of the rural environment; b) product quality and new uses for farm products; c) socio-economic aspects and actions for regions lagging in development. Each of these main headings is divided into six or seven subheadings, comprising a specific research agenda which is now being implemented through assessment and (for some) approval of tendered projects from researchers throughout the Community.

The agricultural part of the FAST programme, consisting of a number of research projects commissioned over the mid-1980s, focused on biotechnology and the food system. Some of the findings of this programme are reported in Traill (1989). The programme did not produce any general framework for the diversity of issues addressed, including food health and safety, irradiation, small firms, catering, and changing consumption patterns. A number of studies have been produced, including edited volumes of seminar and conference proceedings on these and other issues.

Responsible, as it is, for only a tiny part of the total Community research spending in agricultural and food research, the Commission, and through it the EC, cannot hope to direct or control the EC research effort. At the most, the Community programmes can hope to provide a focus on some specific issues arising from market and policy conditions throughout the EC and an encouraging forum for the development of co-operation and communication through the research community, either directly or indirectly. There is little doubt that these limited objectives are being met, at least for that part of the research service which becomes involves in the joint activities. Although those not involved can become isolated from these developments, the Commission places a strong emphasis on dissemination, which with the indirect and informal contact system within the research community, helps considerably to limit this isolation. Nevertheless, current effort and resources devoted to EC-wide research, either in the development of strategies or directly in the conduct of research, falls a long way short of an integrated system, still less integrated with support and intervention policy.

CONCLUSIONS

The interactions between the extraordinary capacity of research scientists and establishments to produce new technologies and the equally remarkable capacity of politicians to interfere with the agricultural sector seem to lend weight to Swift's remark which opened this chapter. The simple economic analysis presented here strongly suggests that it is the policies which lead to most of the problems rather than the pace of technological change and the scientists who produce it. If the incentives exist in favour of producing more food and farming more intensively, then we must expect people to respond to these incentives and to seek out those techniques which allow them to do so more profitably. Technology, according to this argument, is a servant and not a master. Nevertheless, at the farm level, neither politicians nor scientists have managed to repeal the laws of the market-place. The income available to the agricultural sector is determined by market forces, as modified by government intervention. Downward pressure on this total income leads those involved in the industry to seek improvements through a variety of different routes, not least of which is the adoption of new technologies which make more efficient use of resources (land, labour and capital). The consequences of this adoption may exacerbate the pressure on farm incomes, either through the depression of market prices directly or through the continued escalation of policy support costs and the consequent erosion of support prices. The technological treadmill which results from this process may be either benign or malignant. It will be benign so long as the adjustments necessary within the farm sector can be achieved with a minimum of human inconvenience and harm, and so long as the industry continues to produce what the consumer wants, both as safe and high quality food and as an acceptable and attractive countryside. It is malignant if the market pressures result in a failure to match the needs of the producers with the requirements of the consumers.

There is some evidence that the current direction and pace of technological change may be rather more malignant than benign. If markets continue to be distorted through CAP policy intervention, then the signals to producers and consumers are likely to produce 'wrong' responses. Correction of these responses involves either the corrections of the distortions or the introduction of additional restrictions on the use of available technologies. However, the more restrictions are introduced, the more people will quarrel about their relevance or justice, the more likely it is that there will be unintended and unwanted side-effects, and the more effort will be devoted to avoiding or changing them - resources which could be put to better use.

The question of how a different set of market signals and associated policies might produce answers to the three sets of problems posed at the beginning of this chapter warrants another chapter. However, some indications can and should be given in conclusion. Surplus production and despoliation of the rural environment, BST and milk production, and growth hormones in meats were suggested as illustrative of the problems associated with R&D and policy signals. If agricultural support were provided to producers rather than to products (see Chapter 17), and non-marketed environmental goods (see Chapter 8) were encouraged through direct payments to the producers of those goods, then the market would dictate how much production was required of each, and the incentives would be to produce these as cheaply as possible and so earn as much income as possible. Technologies which encouraged the more efficient (lower-cost) production of these goods and services would be encouraged and would be generally beneficial, notwithstanding the lack of a systematic and embracing Community research programme.

However, such a system might well lead to a disparity between producers' views of 'good' and consumers' views, as with BST and growth hormones. Even given the general requirement of governments to ensure that products comply with health and safety standards, there will always remain a problem in that people's views of appropriate standards will differ. Consumers are entitled to know what they are buying, which is an implicit assumption of the market analysis carried out by economists. Providing consumers are appropriately informed about the content and the attendant risk of their food products, it is possible for them to choose whether or not to buy those products in preference to others on offer. Consumer information, in this view, becomes critical and perhaps worthy of public research, support and funding. R&D would be devoted to producing goods and services to meet consumer requirements and countryside user requirements under this scenario, and would become both more defensible and more likely to meet the objectives of the Treaty of Rome than under current policies.

Such a brief outline of an alternative raises more questions than it answers. However, it does emphasise the potential importance of R&D in meeting the policy (and, one must presume, the social) objectives of agricultural policy. But this is not the direction in which the CAP has developed in the past, and there is little evidence that the policy is suddenly about to change direction now. R&D policy, far from being central to the policy development, has remained peripheral and almost totally under national control, in spite of token, if well-meant, attempts to provide some European leadership and integration through the European Commission's efforts. New technologies are potentially valuable in enabling the agricultural sector to adjust to new priorities and demands, but the realisation of this potential requires both the investment in the underlying R&D and the alteration of policy incentives to encourage adoption. There is some reason to be concerned that neither of these conditions will be met.

REFERENCES

AFRC (1988) *Annual Report, 1987/88*, p 5, Swindon.

Cochrane, W.W. (1958) *Farm Prices, Myth and Reality*, University of Minnesota Press, Minneapolis.

European Commission (1988) *Partners in Progress: Coordinated Agricultural Research (1984-88 Programme)*, Official Publications Office of the European Communities, Luxembourg.

European Commission (1990) Community Research Programme in Agriculture, *Green Europe*, 2/90, Official Publications Office of the European Communities, Luxembourg.

Eurostat (1985) *Government Financing of Research and Development, 1975-1984*, Official Publications Office of the European Communities, Luxembourg.

Harvey, D.R. (1988) Research Priorities in Agriculture, *Journal of Agricultural Economics*, Vol. 39, No. 1. pp 81-97.

Lindner, R.K. and Jarrett, F.G. (1978) Supply Shifts and the Size of Research Benefits, *American Journal of Agricultural Economics*, Vol. 60, No. 1, pp 48-58.

McInerney, J.P. (1984) Technology Change as an Instrument of Agricultural Development, *Journal of Agricultural Economics*, Vol. 35 No. 3, pp 379-385.

Schmitz, A. and Seckler, D. (1970) Mechanised Agriculture and Social Welfare: the Case of the Tomato Harvester, *American Journal of Agricultural Economics*, Vol. 52, No. 4, pp 569-57.

Traill, B. (ed.) (1989) *Prospects for the European Food System: a Report from the FAST Programme of the Commission of the EC*, Elsevier, London.

Wise,W. (1984) The Shift of Cost Curves and Agricultural Research Benefits, *Journal of Agricultural Economics*, Vol. 35, pp 2-30.

CHAPTER 11

THE CAP AND THE FOOD INDUSTRY

Alan Swinbank and Simon Harris

INTRODUCTION

It is perhaps not stretching the point to claim that farmers produce raw materials for the food industry, for few farm products are sold directly to households; most undergo some form of preservation or transformation before ultimate sale. Thus, in just the same way that agriculture's input industries are bound in to the food chain and are necessarily affected by the changing commercial fortunes of the farm sector brought about by agricultural policy, so too is the food industry. However, in the European Community, the food industry's interest in the CAP extends beyond the traditional concerns of volume and mix of production, because the CAP helps determine relative food prices and in many instances the CAP is administered by and through the food industry.

ECONOMIC IMPORTANCE OF THE FOOD INDUSTRY

In this chapter the term 'the food industry' should be taken to embrace all those economic activities concerned with the transport, processing and storage of food and drink products between the farm gate and the retail outlet. Thus we exclude catering from our purview, for here the effect of the CAP is rather diffuse.

Reliable data showing the economic importance of the food industry are difficult to obtain, but the member states do carry out an annual enquiry into *industrial* activity (that is, excluding services) which involves all manufacturing enterprise with 20 or more employees. Some salient data from the latest available enquiry are presented in Table 11.1. It is particularly unfortunate that this cut-off applies, for it means that major sections of the 'craft' or 'artisan' businesses that characterise food production in many Community countries are excluded. Similarly, in Table 11.1, it should be recognised that the employment figures for food manufacturing exclude the self-employed, whereas the figures given for agriculture *include* the self-employed and small businesses. Even so it is clear that, in the European Community, food is of major economic importance. Indeed, in the UK, the three activities of farming, food manufacture and food distribution make roughly similar contributions to employment and national income (see Table 11.2).

In the future there is likely to be further growth in the relative importance of the food processing, distribution and retailing activities, at the expense of the farmers' share of retail food expenditure. This increased activity beyond the farm gate reflects the changing lifestyles associated with modern society. In particular, the increasing proportion of women working outside the home, and the trend towards casual eating rather than a formal family meal have meant an emphasis on food preparation by the manufacturer, so that the time spent preparing food at home is minimised. Not surprisingly, the result is to enhance the importance of food preparation, packaging and distribution at the expense of on-farm production.

Table 11.1: Numbers Employed in the Food, Drink and Tobacco Industries, and the Share in Gross Value Added of these Industries (1985)

	Engaged in Agriculture, Forestry, Hunting and Fishing ('000)	Employment Food, Drink and Tobacco Manufacture		Food, Drink and Tobacco Share in Gross Value Added (at Factor Cost) (in All Manufacturing) (%)
		('000)	% of All Manufacture Employment	
Belgium*	105	73	11.1	11.9
Denmark	182	72	20.2	22.6
West Germany	1,390	452	6.7	6.3
Greece*	1,037	56	19.0	18.9
Spain	1,766	342	17.6	18.8
France	1,582	375	9.6	10.6
Ireland	169	47	28.8	**
Italy	2,296	247	8.1	9.1
Luxembourg*	7	2	5.6	6.7
Netherlands	250	123	16.8	17.4
Portugal	969	81	13.0	
UK	620	546	12.3	13.6
EC(12)	10,373	2,406	10.5	10.7

*   1983 figures for the Food, Drink and Tobacco Industries.
**   A figure of 9.1% is reported in the original source.
Source: Commission (1987, Table 2.0.1) for Agriculture; Statistical Office
            (1989, Tables 3.3.1 and 3.3.6) for Food Manufacturing.

Table 11.2: Value Added and Employment in the UK's Food Sector

	Contribution to GDP, 1985 (%)	Employment* June 1986 ('000)	(%)
Agriculture	1.8	315	1.3
Food and Drink Manufacturing	2.9	545	2.2
Food Wholesaling	0.7	235	0.1
Food Retailing**	1.8	720	2.9

* In addition to people in paid employment, there are a large number of self-employed workers in these industries. For example, the addition of farmers and their spouses to the figure for agriculture produces a total comparable to that reported in Table 11.1. Employment figures are expressed as a percentage of the employed workforce.
** Includes confectioners and tobacconists.
Source: MAFF, as reported in Slater (1988, p 4).

The proportion of consumer spending on food taken by food manufacturing and distribution reflects the contribution these sectors make before food reaches the consumer, and the associated costs. The convenience factors built into food, ranging from the presentation of raw materials in ready-to-cook forms to the creation of complete meals ready for the microwave oven, require a range of economic activity, generating employment and business profits. Furthermore, the role of distribution, and its associated costs, has become more important as consumers have sought out 'fresher' food presentations, such as chilled foods, where a whole food chain capable of keeping products at a constant temperature from factory to retail display cabinet is necessary. In this dynamic market, food manufacturers and retailers, of course, brand and advertise their wares as they strive to maximise their share of the consumer's purse, and, in the case of manufacturers, their place on the retailer's shelves. These marketing activities contrast sharply with those of the farmer selling into a protected and relatively static market.

ANNEX II OF THE TREATY OF ROME

The body of EC legislation commonly referred to as 'the Common Agricultural Policy' is derived from Articles 38 to 46 of the Treaty establishing the European Economic Community. There 'agricultural products' are defined as 'the products of the soil, of stock farming and of fisheries and products of first-stage processing directly related to these products'. Furthermore, a list of the items subject to these provisions is given in Annex II to the Treaty. Annex II includes, for example, in addition to 'cereals', 'products of the milling industry; malt and starches; gluten; insulin'. Thus these first-stage products of the cereals processing industry are included in the CAP, whereas the second-stage products of the baking, confectionery, pasta and brewing industries are not.

This distinction between first-stage and second-stage processing can, on occasion, be useful; but equally it can be overplayed, for the commercial activities of the EC's food industry are certainly not all split in this distinctive fashion. There is, however, one fundamental difference: within the Commission, Annex II products are the responsibility of the Directorate General for Agriculture (DG VI), whereas non-Annex II products are dealt with by the much less well resourced Foodstuffs Division of the Directorate General for Internal Market and Industrial Affairs (Division B/2 of DG III). With this split responsibility, involving not just the Directorate Generals, but also their respective Commissioners as well, the EC's food industry has repeatedly complained that their interests are not well represented. DG VI is seen by the food industry lobby as a Directorate General for Agriculture which pays scant regard to the interests of their industry (let alone the interests of consumers, taxpayers, the environment or third country producers).

Paradoxically, the Commission's day-to-day detailed management of agricultural markets on its own competence or through the Management Committee Procedure (fixing import levies, export refunds, and so on), which is largely the responsibility of DG VI, has a more profound effect upon the commercial fortunes of food traders and manufacturers than it does on farmers. The Commission's management role is, of course, important in maintaining market prices, which are of prime concern to farmers; but the day-to-day variations in subsidies and levies which sustain market prices are of little direct concern to farmers, whereas they have immediate implications for traders and manufacturers. The food lobby feels strongly that a body (DG VI) which has such power over their industry should at least display more knowledge of, and sympathy for, their interests. The Commission's preoccupation is one of managing markets for bulk CAP commodities, whereas the food industry's concern is the development of markets for manufactured food products; frequently these two concerns appear contradictory.

FIRST-STAGE PROCESSING: MILK AS A CASE STUDY

As noted in the opening paragraph of this chapter, CAP mechanisms are important in determining not only the volume and mix of European Community farm production, but also the level of food prices; and in many instances CAP support measures are effected through the food industry. The dairy sector can be taken as a case in point.

According to calculations of the Milk Marketing Board for England and Wales, by 1984 some 84 per cent of milk production in the EC(10) was delivered to dairies whilst 16 per cent was retained on farms (see Table 11.3); and these proportions remained unchanged in following years despite the introduction of milk quotas. The bulk of the milk retained on farms was used for animal feed or home consumption. Apart from farmhouse butter-makers in Belgium, farmhouse cheese-makers in the UK (not reflected in the figures in Table 11.3) and other member states, and farmers who sell liquid milk at retail, dairy farmers are in the main dependent upon sales of their raw milk to dairies for their revenue.

Although the aim of the CAP policy mechanisms for milk is to generate a return to farmers which approaches an annually determined *target* price, there are in fact no CAP mechanisms which directly support the price of raw milk. However, import levies and export refunds are important mechanisms for maintaining the market prices of processed products; and in the past intervention buying of butter and skim-milk powder has been a significant source of market support. The price of the farm product (raw milk) is not supported by the CAP, rather it is the price of the processed dairy product which is, and the policy implicitly assumes that the benefits of such price support will be passed back to producers. In fact, the European dairy industry is characterised by its co-operative structure. Dairy farmers are usually owners, through their co-operative, of the processing capacity; and thus as long as the co-operative remains efficient they can expect to benefit from the CAP's largess. However, if milk producers in remote areas are faced with a localised monopsony buyer, as could conceivably be the case in the UK if its Milk Marketing Boards were disbanded, then under present CAP arrangements farmers would have little bargaining strength.

In other product sectors, perhaps reflecting different ownership structures in first-stage processing, CAP arrangements can be different. Thus, in sugar processing, the sugar-beet factory is obliged by EC law to pay a minimum price to growers. Similarly, for a range of crops such as oilseeds and pulses, where EC border protection is slight and farm support is effected by paying a subsidy to the *processor* in the expectation that the financial benefit of the subsidy will be reflected in higher farmgate prices, a minimum producer price will often be set. If the minimum producer price is not paid by the processor, the subsidy in turn is not payable; but this can lead to difficulties, and a farmer may not be in a position to insist upon his rights. For peas and beans, for example, it has been suggested that on occasions animal feed compounders have been unable, or unwilling, to pay the stipulated minimum price because of weak market prices and an inadequate incorporation subsidy: 'market movement has been dependent upon trade taking place at prices effectively below the minimum price, albeit that this had to be disguised'. In the absence of intervention arrangements for these products, 'sticking to the letter of the regulation and having no sales below the minimum price could mean that growers would be severely penalised by the rule that is intended to protect them' (Abbott 1990, p 73).

Table 11.3: Disposals of Milk

Country	% of Total Production		% of Total Farm Utilisation					
	Delivered to Dairies	Retained on Farm	Farm Family	Liquid Sales	Butter Manuf.	Cheese Manuf.	Fed to Livestock	Other
Belgium	74	26	4	7	37	1	50	1
Denmark	94	6	23	-	-	-	77	-
West Germany	92	8	22	11	2	-	65	-
Greece*	46	54	23	11	2	**34	30	-
France	75	25	4	6	3	3	83	1
Ireland	87	13	16	1	1	-	82	-
Italy	72	28	13	35	3	20	29	-
Luxembourg	92	8	12	-	-	-	88	-
Netherlands	98	2	41	-	-	23	36	-
UK	88	12	5	7	1	-	86	1
EC(10)	84	16	10	11	4	7	67	1

Note: 97 per cent of the production recorded is that of cows' milk, but goats' and sheeps' milk is also included.

*   Only one-third of which is cows' milk.

**   Estimate.

Source: Adapted from Tables 28 and 29 of Milk Marketing Board (1986).

For the dairy sector, the main mechanisms for supporting market prices in an over-supplied market have been intervention buying of butter and skim-milk powder, and subsidised exports. Until the imposition of quotas on milk production in 1984, the excess of EC production over consumption was expanding rapidly, leading to substantial costs to the EC's budget. The Commission, through its market management activities, however, can influence very markedly the form an EC 'surplus' takes. The surpluses in the sugar sector, for example, are substantial, and the CAP for sugar provides for automatic intervention purchases of sugar in much the same way as the CAP for milk products used to provide for automatic intervention purchases of butter and skim-milk powder. Despite this, the sugar 'surplus' has never materialised in intervention stocks, whereas the butter 'surplus' frequently did. In large part this was because the Commission, through a weekly export tender, is able to sell subsidised sugar on to world markets in an orderly fashion, whereas large export sales of bulk dairy products have always been more problematic. A second and rather important factor is that, prior to subsidised export, the storage costs of the EC's privately held sugar stocks are met from the EC's budget, with these budget expenditures in turn matched by revenues generated by storage levies charged on the production of all Community sugar, and ultimately paid for by the consumer.

With an over-supplied world market, however, EC milk surpluses were frequently diverted to intervention; indeed certain factories produced principally for intervention. The dilemma for factory managers was clear: if your farmers are producing ever larger quantities of milk that the market cannot readily absorb, then more milk must be turned into butter and skim-milk powder for sale to intervention if the EC's price guarantees are to be realised; and this in turn may mean investment in new butter and skim-milk powder manufacturing capacity. A private company might be unwilling to undertake such a politically risky investment; a co-operative would be less able to resist. Indeed, a justification that is sometimes advanced for the purchase of Unigate's factories by the UK's Milk Marketing Board for England and Wales in 1979 was the MMB's fear that the private sector would be unwilling to maintain a sufficiently large processing capacity to cope with farm production.

Subsequently, the imposition of milk quotas in 1984 and further reductions in the level of quota have rendered much of that butter manufacturing capacity redundant. First-stage processors benefited from the expansion of the CAP in the 1970s and early 1980s, but suffered uncompensated cut-backs from 1984 on. A knock-on effect of the dairy 'reforms', and similar measures in other sectors, was that the cold storage industry, which itself provided a service to the EC's Intervention Agencies, also faced over-capacity. Intervention (and private storage) stocks of butter, which had stood at 1.3 million tonnes at the 1986 year end, had dwindled to less than 0.2 million by the end of 1988 for EC(9) (Milk Marketing Board 1989, p 63).

Throughout the 1980s, EC milk and milk product prices remained well above world market prices and, despite the quota-induced cut-backs, EC milk production continued to exceed consumption by a sizeable margin. Paradoxically, by the late 1980s certain sectors of the EC's dairy industry were reporting a milk 'shortage'. In the absence of quota transfers, all factories had been affected in a similar fashion by the quota-induced supply shortfalls. Butter-making factory managers, faced with weakened price support measures for butter, saw this as evidence of excess capacity in butter manufacture; whereas managers of factories producing products with buoyant EC demand, or selling into profitable export markets on the back of ample EC export subsidies, bemoaned the 'shortage' of milk, induced by quota rigidities, which limited the throughput of their plant. An export-based strategy is not, however, without its risks, for export refunds which today guarantee profits could be reduced tomorrow. McClumpha (1989, p262), of Nestlé, has complained that the 'regular export business of added value products to meet genuine consumer demand, has been subject to sometimes capricious change in export (refunds) interfering with orderly long term market development and promotion.'

SECOND-STAGE PROCESSING

A biscuit-maker would be a typical second-stage processor. Flour, sugar and milk powder - all CAP products - would have to be purchased at EC-supported prices. Butter might be available at reduced prices from intervention stores or new production; but the price and availability of vegetable oils would be little affected by the CAP. [1] If 'subsidised' butter was obtained, a series of checks and controls would have to be introduced to satisfy the authorities that the butter was used for the stated purposes, and had not been fraudulently diverted to the full-price market.

A major part of the raw materials required would only be available at EC, and not world, market prices. This simple fact would render second-stage food manufacturing in the Community unprofitable if it were not protected from manufacturers elsewhere with access to lower-priced raw materials: thus the whole protective mechanism of the CAP and in particular its import levies have to be applied to imported processed foodstuffs. The arrangements are complex and perhaps too inflexible, but they essentially involve a three-monthly review of the import levy for each qualifying food product. The levy consists of two parts: a fixed and a variable component. The fixed component, as its name implies, is an *ad valorem* import tariff designed to protect the EC's

1   It is of course the EC's intention to renegotiate its import arrangements for oilseeds and oilseed products in the context of the GATT negotiations which are due to be concluded in December 1990. Whether other GATT signatories will agree to such a 'rebalancing' of the CAP is highly doubtful. In any event, following the ruling of a GATT Panel in late 1989, the EC has agreed to change its support arrangements for oilseeds.

food processing industry; whereas the variable component is determined by taking the average import levy on each of the product's CAP ingredients over the previous quarter and, on the basis of a notional recipe, fixing an appropriate amount for the product concerned. [2]

Similarly, it is argued, if processed food products are to be exported from the EC, then they must be able to benefit from export refunds of a comparable value to those granted on CAP products. This is rather more problematic, for the present GATT rules prohibit the granting of export subsidies on *processed* products and in May 1983 a GATT Panel of Inquiry ruled that pasta was a processed food product and thus that EC export subsidies were being granted in breach of GATT rules. Despite this, and reflecting the GATT's weak powers, export refunds continue to be paid. Whether or not the present round of GATT negotiations will resolve (or even address) this issue remains to be seen. One possible 'solution' for the processed products issue (if the EC were forced to cease granting export refunds on processed products) would be to grant food manufacturers production refunds on CAP raw materials where used to manufacture products for export. Similar mechanisms have been adopted to try to ensure that EC-based companies have access to CAP raw materials at world market prices for non-food uses (in particular sugar and starch for the chemical industries) where the finished product cannot itself benefit from the CAP's protective mechanisms.

Raw material prices also vary from one member state to another because of the EC's 'green money', and the system of border taxes and subsidies on intra-Community trade known as monetary compensatory amounts (MCAs). [3] Thus, by extension, MCAs have to date applied to many processed food products in much the same way as they apply to CAP commodities. Determining an appropriate MCA for a particular processed product involves the same complexities as are involved in determining the import levy or export refund applied to that product.

Together with common pricing and financial solidarity, Community preference is often taken to be one of the three basic principles, or pillars, of the CAP. Community preference embodies the idea that EC production should have a preferential outlet on the EC's market, at the expense of imported products, and - the food industry would claim - regardless of quality or other characteristics. The mechanism that is usually deployed to ensure Community preference is to fix a threshold (or minimum entry) price at a level in excess of the intervention price. Provided this price difference exceeds the transport costs between surplus and deficit regions, and any quality premium the imported product could command, then imports should only be commercially feasible whilst the Community has a net import requirement. If imports still

2   For further details see Harris *et al.*(1983, pp 243-244).

3   A simple explanation of 'green money' is provided in Chapter 1 and the issue is explored further in Chapters 3 and 9. See also Harris *et al.* (1983, Chapter 8) and Swinbank (1988).

occur when the EC has a net export surplus (unless they are imported on concessional terms) then this is usually taken by CAP apologists as prima-facie evidence of an insufficient level of Community preference. Their knee-jerk reaction would be to seek to increase the gap between threshold and intervention prices. None the less, despite repeated attempts to price imported hard wheats, long-grain rice and other products out of the EC market, consumer preference and/or manufacturing requirements still ensure that sizeable quantities are imported.

AN EXPORT ORIENTATION?

On a number of occasions in this chapter, reference has been made to the real or imaginary constraints the EC's food industry feels it faces on international markets. There is a widespread view, certainly among second-stage processors, that international niceties and GATT regulations have limited the sales of processed, 'value-added', products on world markets. Instead of exporting bulk commodities, the food industry would like to see greater export volumes of processed products, which, they argue, would aid employment and the balance of payments (and of course enhance their profits).

Their view is that world markets for primary agricultural products show relatively slow growth and are subject to alternating periods of 'glut' and 'famine', depending upon the vagaries of weather-induced fluctuations in production and governments' disposal programmes. They are cut-price markets, subject to trade wars as the major exporters jostle for market share. In contrast, the food manufacturers maintain that markets for manufactured food products show strong growth and higher prices. Here manufacturers can create and sustain markets by their use of brands and advertising, thereby maintaining margins and, indeed, setting prices in a way that is not possible for primary commodities.

Notwithstanding the economic arguments deployed, there could be international political objections to an expansion of the EC's exports of processed products. Handy and MacDonald (1989, p 1253) have, for example, reported:

'According to unpublished statistics provided by the Foreign Agricultural Service of USDA, the United States accounts for 21% of world exports in bulk agricultural products but only 5% of world exports in consumer oriented processed food products. Moreover, since the late 1970s, the United States has been running relatively large trade deficits, of $5 to $6 billion annually, in consumer oriented processed food products, while the European Community has shifted from trade deficits to trade surpluses, of around $2 billion annually, in those products.'

Given the EC's tarnished reputation in international agricultural circles, it is difficult to see that the export aspirations of the EC's food industry can be met without a radical restructuring of the CAP, eliminating the need for export subsidies.

CAP or CFP?

The focus of Community policy making in the agricultural and food sectors has been the Common Agricultural Policy since the EEC's inception in 1958. [4] There have been various stages in this process, which could be characterised as follows:

- devising and putting in place the CAP (to 1968),
- operating an open-ended support mechanism (to 1984),
- reform of the CAP, reducing support levels and adapting its open-ended nature (1984 to date).

Nevertheless, despite this evolution, the CAP remains, as it has always been, an agriculturally oriented set of policies.

This constant focus on agriculture and its problems, however, has ignored a fundamental structural change in the European economy: the decline in the importance of agriculture, on the one hand, and the rise in the relative importance of the food manufacturing, distribution and retailing sectors, on the other. It is this economic change, allied with a political change as consumers have become of greater political significance, which has underlain the calls for a common food policy. [5] The excesses of the CAP have allied consumers, outraged at having to pay unnecessarily high prices for unwanted surpluses, with 'free traders' concerned to reduce the trade-distorting effects of EC policies. Both camps can unite in a demand for a CFP which is consumer-oriented; and the food industry lobby (or at least second-stage processors) has found it politic to be associated with such appeals.

Whether or not consumers and food manufacturers are natural allies is a moot point: the food industry lobby has itself often faced internal divisions as some sectors (the first-stage processors) have seen their interests tied in with farming and a continuation of the CAP, whilst other sectors (the second-stage processors) have espoused free trade, at least in raw materials. These divergent interests help explain the relatively weak organisational structure of the CIAA [6] (Confédération des Industries Agro-alimentaires), the food industry's equivalent of the farmers' Brussels-based lobby, COPA (Comité des

---

4   Food law harmonisation and completion of the internal market have, however, been a continuing preoccupation of the Community since the early 1960s, and the '1992' programme has emphasised this importance.

5   Presumably the acronym for a Common Food Policy would be CFP, except that this has been appropriated by the Common Fisheries Policy.

6   On this see Harris (1989, pp 302-303).

Organisations Professionelles Agricoles). And, in turn, this inability or unwillingness to present a united front weakens the food industry's case for a CFP, and for greater recognition of the industry's needs.

'1992'

Discussion of the effects of the Community's single-market programme - under which an internal market 'without internal frontiers in which the free movement of goods, persons, services and capital is ensured', and due to be completed by December 31, 1992 - for the EC's farm and food sectors has been limited. [7] For agriculture this is perhaps not too surprising as the existence of a common Community policy for agriculture, for more than 20 years, has meant that the '1992' programme did not have much left to cover. Even so, the few agricultural issues raised by '1992' are contentious in themselves, and of considerable importance to the food industry: for example, the question whether or not MCAs can be abolished; whether national production quotas should be abolished; and the prospects for a successful harmonisation of veterinary and phytosanitary provisions.

'1992' will be important for the food industry *per se*, but in addition '1992'-induced changes in the food industry will constitute, in time, the principal impact of '1992' for agriculture as well. For the food industry, it is not so much the detail of what is contained in the '1992' programme that is of importance, but rather the change in industrialists' patterns of thought. [8] Because businessmen have been bombarded by governments with '1992' propaganda, telling them that the single market is coming and that they must respond, they have started to take the message on board and to adjust their behaviour. It will be the actions that businessmen take that will make a reality of the politicians' rhetoric. At the food manufacturing level, the principal changes will be a rationalisation of production facilities and a concentration of ownership.

Production facilities will be rationalised as companies having plant in more than one member state, made necessary when each market was thought of as a separate entity, reorganise so that a single larger plant serves a region of the Community, comprising two or more member states, or even the entire Community market. This could lead to abrupt changes in demand for agricultural raw materials as plants in some countries are shut down while those in other countries are expanded. The structural concentration of ownership is also a major concern, as previously national companies seek to

---

7   See, however, Swinbank (1990) and other articles in this special edition of *Food Policy*.

8   The detailed arrangements concerning food law and the concept of mutual recognition are not discussed here in a book on the CAP. For further details see Swinbank (1990).

organise themselves on a Community-wide scale. A rash of takeovers marked the food industry in the closing years of the 1980s, as manufacturers jockeyed for market share by buying up their equivalents in other member states. The result is likely be a repetition, at a Community level, of the structural concentration seen already in several national markets. [9]

Another feature of food manufacturers' behaviour resulting from '1992' will be increased attempts to adopt single-purchasing and, if possible, sourcing strategies covering plants in more than one country. The implication is that, where possible, manufacturers will buy agricultural raw materials where they are cheapest in the Community. The harmonisation of transport provisions under the '1992' programme will make such strategies more feasible.

At the retailer level the effects of '1992' are likely to be slower in showing through. This is essentially because of the deep-seated differences in national cultures and tastes, and hence the difficulty of transplanting a successful retailing concept from one country to another. Nevertheless, some concentration of ownership does seem likely. Furthermore, retailing chains are beginning to form alliances, or buying groups, across national frontiers which will have more muscle than their individual company members, and which may in time develop joint 'own-label' brands.

What is clear is that '1992' should not be treated as a date, but as a process. The changes triggered by the single market, and all that goes with it, will indeed speed up the evolution of the Community's food industry but they will take time to work through. This is a process which will last throughout the 1990s and, probably, a good deal longer.

A further point to note is that the Single European Act extended Community competence to embrace environmental matters. EC environmental legislation will be of increasing importance to farmers, food manufacturers and food distributors in years to come.

CONCLUSIONS

The Community's food industry is a disparate group of businesses in terms of their size, their ownership and their links with agriculture. Collectively they form one of the most important economic sectors of the European Community with sizeable contributions to employment, national income, foreign trade and the price of food. Farmers (and their input suppliers), food manufacturers and food retailers make up a complex and interdependent food chain which is very characteristic of modern society. Farmers do not produce food, but raw materials: they are but one cog in a complex food system. Any attempt to

---

9   As Holmes (1989, p 533) cautions, however, 'The aim of public policy must be to facilitate the realization of scale economies without reducing competition too much.'

mould part of the the the chain, such as the CAP, is bound to influence the whole; and, in the CAP, policy-makers devised a series of price support mechanisms which can only be deployed through the food industry. The CAP is complex and bureaucratic; and any trader or food company that fails to understand the subtleties of policy, have timely access to accurate information, or predict the effect of changes in MCAs, import levies or export refunds can face severe financial penalties. The risks associated with trade, even trade between member states in the absence of the single market of '1992', and the demands on management's time a thorough understanding of the CAP requires are bound to disadvantage the small company in competition with the large. Given that the CAP so strongly influences the commercial fortunes of food companies, the price of food to consumers, the environment, and so on, it is surprising that it is still seen in many quarters as a purely agricultural concern on which farmers and *their* ministers can adjudicate in splendid isolation.

REFERENCES

Abbot, M. (1990) *Combinable Crops and the EEC*, A National Farmers' Union Handbook, BSP Professional Books, Oxford.

Commission of the European Communities (1987) *The Agricultural Situation in the Community: 1986 Report*, Office for Official Publications of the European Communities, Luxembourg.

Handy, C. and MacDonald, J.M. (1989) Multinational Structures and Strategies of US Food Firms, *American Journal of Agricultural Economics*, Vol. 71, No. 5, pp 1246-1254.

Harris, S.A. (1989) Agricultural Policy and its Implications for Food Marketing Functions. In: Spedding, C.R.W. (ed) *The Human Food Chain*, Elsevier Applied Science, London.

Harris, S.A., Swinbank, A, and Wilkinson, G.A. (1983) *The Food and Farm Policies of the European Community*, John Wiley, Chichester.

Holmes, P. (1989) Economies of Scale, Expectations and Europe 1992, *World Economy*, 12(4), pp 525-37.

McClumpha, A.D. (1989) International Trade Implications. In: Spedding, C.R.W. (ed) *The Human Food Chain*, Elsevier Applied Science, London.

Milk Marketing Board for England and Wales (1986) *EEC Dairy Facts and Figures 1986*, MMB, Thames Ditton.

Milk Marketing Board for England and Wales (1989) *EEC Dairy Facts and Figures 1989*, MMB, Thames Ditton.

Slater, J.M. (1988) The Food Sector in the UK. In: Burns, J.A. and Swinbank, A. (eds) *Competition Policy in the Food Industries*, Food Economics Study No. 4, Department of Agricultural Economics and Management, University of Reading.

Statistical Office of the European Communities (1989) *Industry Statistical Yearbook 1988,* Office for Official Publications of the European Communities, Luxembourg.

Swinbank, A. (1988) Green 'Money', MCAs and the Green ECU: Policy Contortions in the 1980s, *Agra-Europe* Special Report No. 47, Agra-Europe, Tunbridge Wells.

Swinbank, A. (1990), Implications of 1992 for EEC Farm and Food Policies, *Food Policy,* 15(2), pp 102-110.

PART III

THE CAP AND THE WORLD

CHAPTER 12

THE CAP AND WORLD TRADE

Allan Buckwell

INTRODUCTION

This chapter attempts an explanation of the nature and causes of the distortive effects of the CAP on world trade in agricultural produce, culminating in an assessment of their role in the Uruguay round of negotiations under the General Agreement on Tariffs and Trade (GATT). The order in which the arguments are made is first to show the nature of the connections between the CAP and world trade. Second, some statistics are presented which describe current European trade in agricultural products and its evolution over the last decade or so. Third is a discussion of the interactions between commodities and countries, which have created the tangled web of world agricultural trade. This situation was described as long ago as 1973 by Gale Johnson in his book *World Agriculture in Disarray*. The situation has, if anything, worsened since that time. Fourth, attention is focused on the current GATT round, indicating some of the major issues and the stances of the main actors in these negotiations. Next, although quantitatively less developed countries (LDCs) are of much smaller significance in world trade in temperate zone products, it would be remiss not to say something about the interrelationship between the CAP and LDCs, in this general chapter on the CAP and world trade. Finally, these thoughts are drawn together with some concluding remarks.

THE CONNECTION BETWEEN THE CAP AND WORLD TRADE

That there would be a significant link between the domestic farm policy of one of the most important trading blocs in the world and world trade was guaranteed by the way in which the objectives of the CAP were defined in 1957 under Article 39 of the Treaty of Rome, already referred to several times in this book. Briefly, these were to increase farm productivity, thereby to ensure a fair standard of living for those engaged in agriculture. It was also intended to achieve stability in agriculture, security of food supplies and last, and as it turned out least, to ensure fair prices for consumers.

The policy took about a decade to develop. It was not until the late 1960s that a common policy was achieved for farm products for the original six countries and even then there is some debate about whether it was achieved in practice. Almost immediately, the Community found itself in a process of enlargement. Britain, Denmark and Ireland joined in 1973, Greece joined in 1981, and the most recent expansion took place with Spain and Portugal in 1988.

The essence of the Common Agricultural Policy is to support agriculture primarily through price support using a bewildering variety of instruments. However, it is possible to simplify these to explain the impacts of the CAP on trade. This was done in Chapter 1, and the same ideas are demonstrated in Figures 12.1 to 12.4 using the elementary microeconomic constructs of supply and demand curves.

Figure 12.1 depicts the market for an agricultural product, cereals say, for the whole Community. Quantities produced, consumed and traded in millions of tonnes are shown on the horizontal axis; prices, measured in ECU/tonne, are shown on the vertical axis. It is conventional to show an upward-sloping (or positive) relationship between the quantities supplied and price, and a downward-sloping (or negative) relationship between the quantities demanded and price. If the EC market were completely isolated from the rest of the world, it would find an equilibrium quantity Qe, at which the volume of cereals produced and consumed were roughly in balance. Of course, the European Community is not totally isolated from the rest of the world. The EC exists as part of a global trading system. Indeed, it was largely through the efforts of European citizens down the ages that this system evolved.

Figure 12.1: European Autarchy

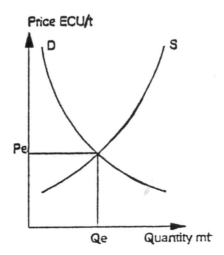

Practical experience over the last century demonstrates that cereal prices outside the Community would be below the equilibrium price which would occur inside the Community if it were isolated as depicted in Figure 12.1. This is probably also true for most other temperate products in Europe with the possible exception of wine.

To analyse the effect of moving from this situation of autarchy to free trade it is convenient to make the simplifying assumption that the European Community could buy as much as it liked of agricultural commodities from the world market without affecting their prices. Whilst this is unlikely to be true, it does not do substantial damage to the analytical conclusions drawn below. The assumption enables the world cereal supply to be depicted in Figure 12.2 as a horizontal straight line at the world price Pw. This suggests that the EC can purchase any volume of cereals from the world market it chooses at that price. In this situation the quantity that would be offered for sale by Community farmers would be the amount, Qp. This would clearly be a lot less than the quantity the consumers would wish to utilise at such low world prices, Qc. The difference would be made up by imports. Thus the presumption is made that with unrestrained markets the Community would be a net cereals importer. This is also expected to be the case for most other agricultural commodities.

Figure 12.2: Free Trade at World Prices

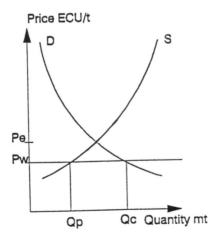

This situation of free agricultural trade does not, of course, exist. It has never existed for the major continental countries in Europe and not even for the UK for a very long time. It has been judged that it is politically quite unacceptable to expose Community farmers to prices as low (and unstable) as those on the world market. The ruling situation on European markets observed for a very long time is that of protection. The simplest form of protection, particularly starting from a net import situation, is to have some kind of import barriers. This could be achieved by setting a threshold price, Pt, above the world price, and insisting that no cereals are to enter the community market below this price. Such a policy, depicted in Figure 12.3, is implemented by imposing a border tax to make up the difference between the threshold price (Pt) and the world price (Pw). This is one of the principal means by which the Community has protected its internal market. It can be shown to have five effects. First, with the higher and more stable internal price, it encourages additional production in the Community. Second, it discourages consumption of Community cereals. Third, it reduces the volume of imports because of these two effects. Fourth, it raises revenues as long as some imports remain. These are collected as import taxes or, more accurately, variable import levies. They accrue as part of the Own Resources to the budget of the European Community. The fifth consequence of border protection using a threshold price mechanism is that the Community market is insulated against fluctuations in the world price. As the world market price rises and falls because of fluctuations in world grain production and use, the gap between the constant threshold price and the varying world price must vary too. This gap is the variable import levy.

Figure 12.3: Protected Market - Imports Remain

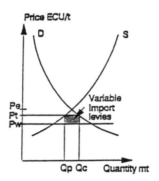

The picture depicted in Figure 12.3 roughly approximates to the current regime for maize (that is, corn), where the EC remains a significant net importer. However, it no longer describes the situation for many other commodities, because of two changes which have taken place. First, the threshold price of grain has been raised compared with the world price. Second, the supply curve has been moving downwards and to the right as a result of cost-reducing technical and structural change in European agriculture. The consequence of these tendencies has been radically to alter the situation, as shown in Figure 12.4.

The new internal price level is referred to as an intervention price, Pi, which is considerably above the new Community equilibrium price Pe where the EC demand curve intersects the new rightward-shifted supply curve S'. It is also a long way above the world price level. The Community is thus now in the situation where production, Qp, exceeds consumption and has switched from being a net importer of cereals as in Figure 12.3 to being a net exporter as in Figure 12.4.

A major problem for the Community is to find countries willing to buy EC cereal exports at Community prices. Few countries would wish to do this if they had the opportunity to buy grain at the world price. Therefore, the EC has to subsidise exports by the difference between the internal and world price.

Figure 12.4: Protected Market - with Subsidised Exports and Supply Shift

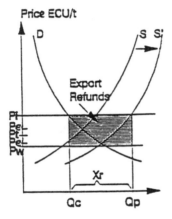

This is the export refund shown in Figure 12.4. If this amount of refund is paid on the volume of exports shown Xr, it involves a substantial cost to the EC budget. This is indicated by the shaded area. It is not difficult to imagine that, as rightward shifts in the supply curve continue, and as expanding EC exports themselves depress world market prices, the budgetary cost will escalate. This is precisely what has happened. It is a cause of the explosion in budgetary costs which drove the CAP to successive crisis points during the 1980s.

The protection given by the intervention price and the export refund system is not the only way EC agriculture is protected. There exists a battery of other measures both under the CAP and in national legislation in the member states. These include tax allowances, capital grants and subsidies for research and development. Many of these have the effect of shifting the supply curve to the right. [1] They therefore expand the volume of produce available for exports and thus intensify the impacts on the rest of the world.

---

1   The various implications of the shift in supply curve due to government's expenditure on agricultural research and development were analysed in detail in Chapter 10.

This explains the mechanism through which the European Community has transformed itself from being a net importer of most agricultural commodities to a net exporter. Not surprisingly, traditional exporters of agricultural goods do not feel happy about this transformation. The other main feature of the EC support arrangements is that it is a very effective way of insulating the European Community from world markets. Whenever world prices go up or down, reflecting respectively shortage or surplus outside the EC, little of this information is transmitted to Community farmers or users of agricultural commodities. Likewise, fluctuations in the balance of production and utilisation within the EC are exported to the rest of the world, destabilising the market in the process. Thus the main complaints by countries from the rest of the world against the European Community are the loss of markets and the destabilisation of world markets.

THE TRANSFORMATION OF THE EC TRADING STANCE

Figures 12.5 to 12.8 and Table 12.1 show the transformation of EC agriculture from a situation of overall net imports of agricultural produce to a position of net exports over the decade 1975 to 1986 for cereals, sugar, wine and beef and veal. Up until 1979/80 the figures refer to the European Community of nine and from 1980/81 onwards they refer to the Community of ten, that is, excluding Spain and Portugal. The picture is very clear. In every case, exports have risen. The time path is erratic but the upward trend is unmistakable. Imports have, over the same period, been static or falling. Thus the balance between the two, the net export balance, has been growing very rapidly. The point at which the Community switched from being a net importer to being a net exporter was 1979 for cereals, it was earlier for sugar, in 1975, for wine it was 1976, and for beef and veal 1979. This transition is seen by the rest of the world as an important effect of the Common Agricultural Policy. Of course it is not just the CAP which has fostered the change but the multitude of Community and domestic supports for agriculture which have encouraged a rapid expansion of production at a much faster rate than growth in utilisation.

This switch in trade position is also documented in Table 12.1. The first two columns of the table show self-supply ratios: the ratio of EC production to EC utilisation. When this ratio is over one hundred it indicates that the EC is a net exporter. For wheat and barley, as early as 1973 and 1974 the EC was already a net exporter. Of major cereals it is only for maize that the Community was and remains a large net importer. The table shows that sheepmeat and vegetable oils are still imported. However, there is a clear trend of increased self-supply for these commodities.

Figure 12.5:   Community Trade in Cereals. EUR12 1973/74 to 1987/88

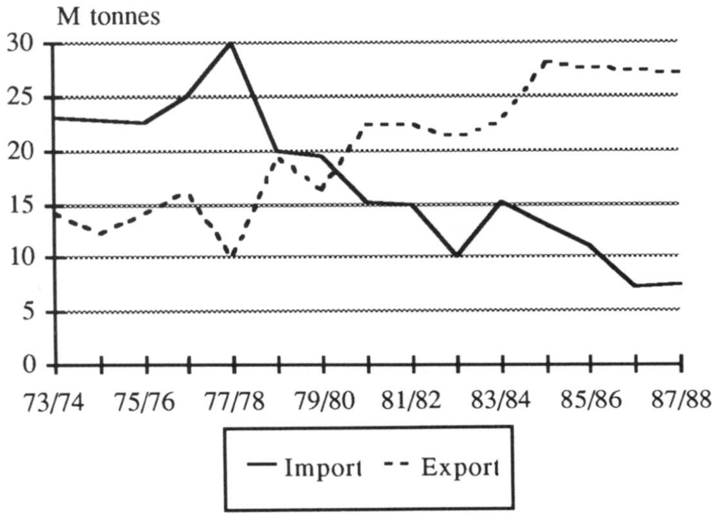

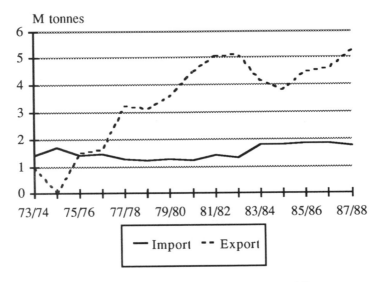

Source: Agricultural Statistics in the Community. Annual Reports.

Figure 12.6:   Community Trade in Sugar. EUR12 1973/74 to 1987/88

Source: Agricultural Statistics in the Community. Annual Reports.

Figure 12.7: Community Trade in Wine. EUR12 1973/74 to 1987/88

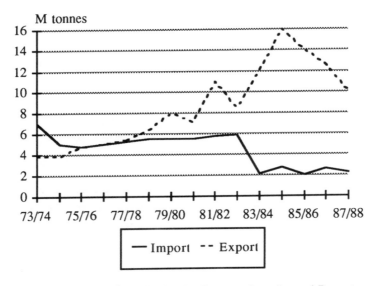

Source: Agricultural Statistics in the Community. Annual Reports.

Figure 12.8: Community Trade in Beef and Veal. EUR12 1973/74 to 1987/88

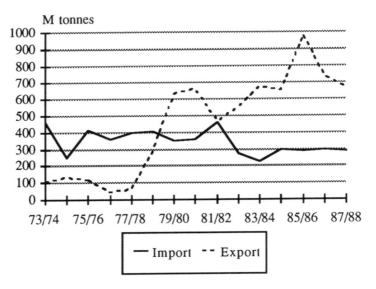

Source: Agricultural Statistics in the Community. Annual Reports.

During this transformation process the Community has become a very large actor on world markets. The last two columns of Table 12.1 show the share of the EC in world trade in each product. Despite the fact that the Community is a net exporter of most products, it is still the largest importer of agricultural produce in the world. Simply because of its sheer economic magnitude, the imports that still remain are substantial; it imports nearly half of vegetable oils traded on the world market and over a quarter of the sheepmeat. Overall, EC imports of agricultural produce are almost a quarter of total world trade in these products. On the export side, the Community has become a very significant exporter, and it is not unreasonable to ask whether it is sensible that the EC, with its relatively high costs of production, should account for between a third and a half of world trade in some commodities. This is particularly so when those exports cost so much to dispose of on to world markets.

This gives some flavour of how the Community interfaces and in some cases distorts world trade, and some idea of the magnitudes involved and how they have changed in the recent past. The CAP has created conditions under which European agriculture has expanded and has become a major actor in world markets. If these changes had come about as a result of the highly productive and thus competitive nature of European agriculture, citizens in Europe and abroad would have little to complain about. However, because the changes have been accompanied by a large and rapidly rising injection of public support, and because EC agricultural product prices are significantly (sometimes orders of magnitude) higher than those in competing countries, there has been a chorus of complaints against the EC and the CAP.

Table 12.1: External Agricultural Trade and the European Community

	Self-supply Ratio			(%) Share in World Trade 1984	
	1973/74	1984/85	1985/86	Imports	Exports
Wheat	104	129	126	3.4	16.7
Barley	105	124	124	-	33.0
Butter	98	134	133	13.1	44.2
Meat	96	102	102	13.9	19.7
Beef	95	108	106	9.7	23.5
Sheepmeat	66	76	80	27.0	-
Vegetable Oils	81	86	56	48.1	0.2
Overall				23.0	12.0

THE TANGLED WEB OF AGRICULTURAL TRADE

The broad outlines of the Community's agricultural trade situation might seem bad enough to those who consider that trade flows should arise as a result of comparative advantage. On closer inspection the situation is even more distorted because of the large imbalances which exist in the degree of protection given to different commodities. This can be illustrated by considering the interactions in production and consumption between sugar, cereals, oilseeds and animal products.

Most of these products are protected, but not all to the same extent. There are significantly different rates of protection offered for different commodities and thus distortions arise in production, consumption and trade. This feature is not restricted to the European Community. It also occurs in the United States and in many other developed and developing countries. These distortions are discussed and analysed in depth in Koester *et al.* (1987). The idea is summarised in Table 12.2.

Table 12.2:   The Tangled Web of World Agricultural Trade.   Uneven Rates of Protection Create Enormous National and International Distortions

EC		US
Sugar	Very high	Sugar
Dairy	protection	Dairy
Cereals		Wheat
Oilseeds		Coarse grains
		Soya
Oils and fats	Low or	Meat
Cereals substitutes	zero	Fruit and vegetables

Source: Koester *et al.* (1988).

In the EC sugar and dairy products are protected at very high levels. Moving down the table, crudely speaking, the commodities listed enjoy lower protection. [2] In fact little or no protection is offered against imports or for domestic production of cereal substitutes, proteins and fruit and vegetables. In the United States there is a similar situation. As a result of these uneven rates of protection, large budgetary and consumer costs are incurred, both in Europe and in North America, and many trade flows are a result of the market distortions rather than real differences in production or demand.

---

2   The hierarchy is much the same as that given in Chapter 7 when considering the differential effect of the CAP on consumer food prices.

232

Concrete examples best exemplify these distortions: high cereal prices in the European Community have stimulated cereals production and discouraged consumption, particularly for animal feed. Animal feed compounders prefer not to purchase high-cost Community grain as a basis of rations for pigs and chickens and cows; they would prefer to take advantage of low-price, energy-rich cereal substitutes such as manioc (tapioca or cassava) or citrus pulp. These have not always been thought of as useful products; they were previously seen as waste or by-products. However, within the Community they are now seen as low-cost sources of energy. They are easily mixed with maize gluten or soya bean, or other protein-rich materials, to create nutritionally balanced animal feed. Partly as a result of previous GATT bindings and partly because these products were not produced in the EC, these feed ingredients enjoy almost no border protection. Thus, over the last decade the EC has imported very large quantities of cereal substitutes and proteins for use in the animal feed industry instead of domestically produced cereals. The displaced European cereals then are exported on to the world market at large budgetary cost to the EC and further depress world market prices.

From the perspective of the other side of the Atlantic, the United States views the depressed state of the world markets as being partly caused by the actions of the European Community. They consider that grain prices are lower than they would be if there were no European support policies. Relatively, this makes soya look better in the US although it is not protected to a great extent. On the other hand, sugar in the US is protected at a very high level, creating a large disincentive to US food manufacturers to use sugar. A consequence has been the development of a large corn syrup industry which provides a very acceptable substitute for sugar (called high-fructose corn syrup) used throughout the soft drinks industry in the United States. The extraction of the fructose from grain leaves a high-protein by-product called maize gluten. This is now an important source of protein in the EC where it is mixed with cheap carbohydrate and fed principally to pigs, poultry and dairy cows. The resulting relatively low feedstock to livestock price ratios contribute towards further expansion of those sectors in the EC. The consequential surpluses of milk powder, butter and meat are then exported with subsidies on to the world market, depressing world markets in these products. This is, naturally enough, resented by traditional exporters of these products such as the United States and New Zealand. The domestic distortions thus spread from one sector to another and from one country to another.

Each commodity spills over into other commodities because of interlinkages in production or consumption or both. Distorting one market creates distortions in the other markets, and this merry-go-round has been in motion for decades. The result is that a significant part of world trade in agricultural products has little to do with the comparative costs of production and much more to do with these distortive policies and the investments and vested interests they stimulate.

The twin problems of massive intervention into agricultural commodity markets and gross distortions of these markets form the background to the discussions under the GATT.

THE URUGUAY ROUND OF GATT NEGOTIATIONS

The GATT was set up after World War II in order to bring about a liberalised regime for world trade in all products. The GATT operates through successive rounds of negotiations. Each round takes three or four years and there are usually several years between rounds. Seven such rounds have taken place since GATT was set up. Over this period, substantial progress has been made in liberalising trade in manufactured goods for which tariffs have been negotiated away to very low levels. One result is that agriculture, which has escaped all attempts at negotiations under GATT, is an area that stands out as being highly protected and distorted with substantial barriers to trade. In previous GATT rounds the EC has always resisted attempts to liberalise agricultural trade. It was able to claim that the CAP was still in its formative stage and could therefore not be the subject of international negotiations. Now that the EC and the CAP have 'matured' and the Community's trade stance has been altered so dramatically, these are no longer acceptable arguments.

As the problems outlined above emerged it was agreed that a serious attempt should be made in the current GATT round to liberalise agricultural markets. Together with trade in services, non-tariff barriers and intellectual property rights, agricultural trade has been elevated to one of the major subjects of the current round. The Uruguay round was launched in 1986 at Punta del Este with a declaration signed by all participants which sought liberalisation of trade in agriculture. Specific reference was made to improving market access, enabling exporters to get into formerly protected markets, and to increasing discipline on the use of all direct and indirect subsidies and other measures affecting directly and indirectly agricultural trade. This was a very clear reference to the fact that negotiations were not to focus primarily on export subsidies, which is what the European Community feared, but that the whole range of protective measures towards agriculture would be considered.

At the time of writing, the four-year negotiations were in their final stages and progress was difficult and slow. In the initial round of proposals tabled in 1987 there was such a wide gulf between the ideas of the principal parties, the United States, the European Community, the Cairn's group (a group of fourteen developed and developing country agricultural exporters led by Australia) and developing countries that it took until April 1989 to find any kind of consensus at all. The United States initially proposed rather extreme measures, often referred to as the 'zero option'. Their suggestion was that participants should agree the long-term objective of abolishing all trade-distorting support to agriculture within a decade. They proposed to allow countries to continue to protect agriculture providing such supports are 'decoupled' from production. To implement such a process of liberalisation the US suggested that attempts be made to devise a method of agreeing a common basis for measuring distortions in agricultural trade, and these measures would then be used to guide the negotiations and to monitor progress towards liberalisation. One suggestion for this measurement framework was the concept of producer consumer subsidy equivalents, PSEs and CSEs respectively. [3] These provide a method of converting all support measures into an equivalent producer or consumer subsidy.

The European Community proposed a different approach. The Community was less concerned with long-term objectives than with short-term measures which would liberalise markets. The Community wished to claim credit for the reductions in farm supports it had made since the introduction of milk quotas in 1984. The Community was also eager to base negotiations on an overall measurement of support (initially called the SMU - Support Measurement Unit and later modified to the AMS - Aggregate Measure of Support) to provide participants with the flexibility to choose their own support methods and the balance of protection between commodities.

The Japanese are another important party to the GATT talks. They have extremely high levels of protection in some sectors of their agriculture such as rice and livestock products. Under external pressure they have already made some progress in liberalising livestock markets but they have not been prepared to cut their support to the rice sector. They argue that rice production is deeply embedded in Japanese culture and history and there are strong social and regional (and thus political) reasons to protect the existing small-scale, mostly part-time, rice producers. The Japanese have been unwilling to put this protection on the negotiating table.

Two other participants to the talks require mention: the developing countries, whose interests are summarised below and discussed at length in Chapter 13, and the Cairn's group. The latter's proposals are a compromise between the US and EC positions: they were in favour of agreeing on long-term objectives of eliminating agricultural support, but they were eager for progress on short-term measures too. They proposed a freeze on protection, and sought an explicit framework for reducing protection over the longer period.

The political momentum towards wholesale reforms of agricultural policies with the aim of liberalising trade in farm produce was undoubtedly helped by the wealth of evidence from academic economists showing the potential benefits of this process. These have been reviewed by Winters (1986) and discussed by Sarris (1991) and Buckwell (1991). To quantify the benefits in a consistent way for the major developed economies, the OECD modelled the effects of trade liberalisation for six agricultural trading countries/regions - Australia, Canada, the EC, Japan, New Zealand and the US. The result showed

---

3   The concepts of producer and consumer subsidy equivalents were originated by Josling in a study for FAO in the mid 1970s. They are discussed in detail in OECD (1987).

that the real income cost of agricultural supports in 1986-88 was just under 1 per cent, valued at $72 billion. Public discussion of these enormous costs of farm supports increased the pressure on the GATT negotiators to resolve the differences in the agricultural negotiations.

By the time of the mid-term review in December 1988 in Montreal, no real progress had been made to resolve the differences in attitudes towards liberalising agricultural trade. When the mid-term review concluded in April 1989, it contained some generalised objectives and guidelines for the remainder of the negotiations which to the outside observer seemed very little further forward than the original Punta del Este declaration. During the autumn of 1989 the US appeared to moderate its original proposals and, as a concrete mechanism, suggested that all quantitative restraints on imports be replaced by equivalent tariffs which could subsequently be negotiated down. The EC, after initial hostility to this idea, seemed to accept at least a partial 'tariffication' but still insisted on the use of aggregate support measures and the need for re-balancing protection.

During the final year (1990) of the negotiations, there has been some evidence of convergence of the major parties. This was only achieved by the intervention of the heads of government of the so-called group of seven (G7) countries who at their Houston summit in July 1990 commended to their negotiators the framework outlined by the Chairman of the GATT Agriculture Negotiating Group, Mr Aart de Zeeuw. The principal elements of the framework dealt with internal support measures, border protection and export competition. To reduce internal supports each party will make commitments to measure, for each commodity, the 1988 support level in which all forms of support are included. These will then be subject to an agreed timetable of 'substantial and progressive' reductions. Exceptions are allowed for certain kinds of direct payments not related to specific commodities and which are taxpayer-funded and divorced from output levels. The proposals on border protection are to convert all border measures into tariff equivalents which will be bound and then cut over a period to be agreed. The suggestions for improving export competition are to agree on commitments to reduce all forms of export assistance.

This chapter is not the appropriate place to assess the likely outcome of the Uruguay round. By the end of 1990, agreement had still not been reached, with the EC's 'offer', of a 30 per cent reduction in aggregate support (itself reached only after a very painful negotiation between member states), rejected by the other parties. The attitude of the EC will clearly be of great importance. To a large extent the pressure on the EC for further wholesale reform of the CAP was lower in 1990 than in most of the 1980s. This resulted from the Brussels Summit in February 1988, which released further budgetary resources and agreed the stabiliser measures to contain the growth in agricultural output. Other factors were the appreciation of the US dollar against the ECU and the upturn in world markets for many products in 1988 and 1989. None the less, it is undeniable that Community farmers have been exposed to steady cuts in real prices and, with the exception of dairy farmers, their incomes have declined correspondingly. There are clear limits to the rate at which EC politicians are prepared to see this process continue. Wholesale abandonment of existing farm supports is not a likely prospect.

Another trade problem between the EC and the US, and which has been referred to GATT for adjudication, concerns the use of the steroid hormones in beef production. The EC has prohibited the use of these substances and has consequently banned the import of beef from countries where they are in widespread use. This has mostly affected the United States. The US claims that, because there is no scientific evidence to justify it, the ban is an illegal anti-trade measure. On the other hand, the Community claims the right to protect its consumers from products they consider produced in an undesirable way. The interest in this issue lies not so much in the value of the affected trade, which is a relatively small volume of low-grade beef destined mostly for pet food, but in the larger principle of legitimate grounds for countries to limit trade. With a growing trend towards green consumerism in the EC, and a distrust of products of what is perceived to be an over-intensive agriculture, the steroid beef ban could be a precedent for similar treatment of other 'undesirable' technologies. This could lead to further serious trade disputes which have their origins in agricultural or food policy.

THE CAP AND DEVELOPING COUNTRIES

There is little doubt that the Common Agricultural Policy has a profound effect on the agricultural sectors of many developing countries. Whether that effect is positive or negative is rather more controversial. These matters are dealt with fully in the next chapter; they are summarised here for completeness. The first point to make is that European agriculture produces temperate zone products which, apart from a few products, are not *directly* in competition with the tropical exports of most LDCs. The important conflicts arise where tropical products are substitutes for temperate products - particularly the case of sugar and tropical oils and fats.

The CAP provides benefits for some developing countries. First, this is because the impact of the CAP is to depress world market prices, which means that developing countries which import grain, meat or dairy products from the world market will pay lower prices than would be the case in the absence of the CAP. This is presumably of some benefit to consumers in these countries although their farmers would have a different view. Economists have tried to quantify these benefits. Second, some LDCs benefit from the side-effects of the distortions in agricultural trade described earlier in this chapter. A prime example is the large volume of manioc which Thailand exports to the European Community. This trade would not exist were it not for the high cereal prices coupled with zero or low duties on cereal substitutes. Third, there are a number of countries who are beneficiaries of trade under the Lomé Agreement, in particular Botswana's beef exports and Caribbean sugar producers. They have preferential access to the European Community for certain volumes of these products and they are paid prices which approximate to European levels rather than the lower world market prices. Fourth, it is argued by some that the existence of large volumes of stocks of agricultural products and the availability of these stocks for famine relief are a benefit under certain circumstances.

However, there are a number of arguments which suggest that the CAP creates difficulties and damage to the economic interests of developing countries. First, low world prices which are reflected on to developing country agricultural markets are a disincentive to production in those countries. This may slow the pace of economic development in some countries which could produce for world grain markets. Second, there is no doubt that the destabilising effect of the CAP hurts developing countries who depend on exports to the world market and have to suffer the low prices and the uncertainties. Third, the lack of access to the European market does affect certain developing countries. Fourth, there are arguments that the macro-economic effects of the CAP and protection in other developed countries harm developing countries. There have been a number of analyses in recent years based on general equilibrium models which take account of the economy-wide resource use and income effects. These studies have shown that, by protecting agriculture, the industrialised countries of Europe and North America have retained resources in agriculture. The result is lower economic growth than would otherwise be the case and, therefore, a lower growth in markets for exports of developing countries. These roundabout arguments have been quantified, for example by the OECD (1990), who calculated that the loss to developing countries, because of this protectionism, could amount to 26 billion dollars per annum.

It is difficult to draw clear conclusions from these arguments. The CAP clearly has significant impacts on some developing countries, some harmful, some beneficial. Generalised conclusions about its impacts on developing countries as a group are not possible but neither are they very useful.

CONCLUSIONS

Five conclusions may be drawn from this discussion of the effects of the Common Agricultural Policy on world markets. First, the international effects of the CAP are largely unintended consequences of a domestic policy. The European Community did not set out to have these effects.

Second, until comparatively recently, the international impacts of the CAP were not widely recognised or acknowledged, at least within the EC. It is only really during the 1980s that the Community has been made aware of the extent and the size of these effects although, of course, the effects themselves have grown.

The third conclusion is that, because the European Community has become such a large actor on the world market - the second largest exporter of agricultural commodities in the world and the largest importer - it can no longer avoid the responsibilities this position brings with it. The EC is a prominent part of the global community which subscribes to an open trading economy. As agriculture accounts for only 5 per cent of EC gross product and 8 per cent of employment, it cannot be argued that it is in the interests of the EC that trade distortions in agriculture should threaten the future of the free trade system. For this reason, the EC was bound to subscribe to the Punta del Este declaration to reduce the protection in agriculture.

The fourth conclusion, therefore, is that the Community feels itself under enormous pressure, both from other developed countries and from developing countries, to curb protection under the CAP.

However, the fifth conclusion is that, because the CAP is such a complex compromise between countries and commodities within Europe, it is extremely difficult to make changes in the CAP. All reforms have to be incremental and take place over a period of years. There is almost no prospect that the Community has the desire or capability to dismantle its agricultural policy. Therefore, it is most unlikely that in the current GATT round there could be substantial or rapid reductions in protection in European agriculture. The Community has tried to show that it is willing to make progress in other areas. For example, it has supported proposals substantially to liberalise trade in tropical products.

Finally, in a chapter contributed in memory of John Ashton, it is wise to refer, as he would have done, to a broad issue potentially of global importance, the developments in Eastern Europe. There is little doubt that the performance of agriculture in Eastern Europe and the Soviet Union is hampered less by natural conditions and technical know-how than by the inefficiencies of their economic system. The economic reforms under way in these countries have the potential to transform their agricultural performance. These developments are in their infancy and are not yet securely established; however, if realised, they could have profound effects on EC agriculture by denying the Community export markets, and possibly even by providing competition through low-price exports of their own.

REFERENCES

Buckwell, A.E. (1991) Problems of Modelling the Effects of Liberalising Agricultural Trade, *European Economic Review* (forthcoming).

Johnson, D.G. (1973) *World Agriculture in Disarray*, Fontana/Collins, London.

Koester, U. *et al* (1987)*Disharmonies in US and EC Agricultural Policy Measures.* European Communities, Brussels.

Loo, E. and Tower, E. (1988) Agricultural Protectionism and the Less Developed Countries: the Relationship Between Agricultural Prices, Debt Servicing Capacities and the Need for Development aid, *Centre for International Economics and the Trade Policy Research Centre,* Seminar 4.5.1988, London.

OECD (1987) Calculations of Producer Subsidy Equivalents (PSEs) and Consumer Subsidy Equivalents (CSEs) for the European Community, Annex 3 of *National Policies and Agricultural Trade: Study on the European Economic Community,* OECD, Paris, pp 244-997.

OECD (1990) *Modelling the Effects of Agricultural Policies,* OECD, Paris.

Sarris, A. (1991) European Agriculture, International Markets and LDC Growth and Food Security, *European Economic Review* (forthcoming).

Winters, L.A. (1986) The Economic Consequences of Agricultural Support: a Survey, OECD *Economic Studies,* No. 9 (Autumn), Paris, pp 7-54.

CHAPTER 13

THE CAP AND ITS EFFECTS ON DEVELOPING COUNTRIES

John Lingard and Lionel Hubbard

INTRODUCTION

The adverse effects of protectionist agricultural policies of rich countries on economic development in poorer countries are now an issue of growing international concern. Harvey (1988) has spoken of a 'global imbalance of food' and 'the moral obscenity of developed country surpluses ... compared with chronic and acute famines in parts of the developing world' - an issue to which he returns in the final chapter of this book. In particular, the effects of the Common Agricultural Policy and the European Community's agricultural trade on both the level and stability of world agricultural prices have come under close scrutiny during the Uruguay round of GATT negotiations.

However, the implications of the CAP for the LDCs are not immediately obvious, as mentioned briefly in the previous chapter. The distorting effects of European protectionism have undoubtedly hit agricultural exports of some LDCs, particularly in commodities like sugar, beef, and cereals; but the impacts are neither universally harmful nor uniform. Some groups of LDCs (food importers) and some groups within LDCs (food consumers) may actually benefit from the trade effects of the Community's agricultural policies. There are different implications for African food importers compared with Latin American net food exporters, and it is not possible a priori to determine what particular CAP changes would unambiguously safeguard LDC's interests. The export markets for LDCs, particularly in temperate products, have been reduced by the EC's external tariff, but world food prices have been lowered by Community exports, which has aided the food-importing low-income countries of Africa and Asia.

Agricultural trade between the EC and LDCs is also influenced by a set of preferential trading relationships, mostly under the auspices of the Lomé Convention. We refer briefly to these towards the end of the chapter. However, in the main the trade concessions granted have not involved products covered by the CAP (unlike the Agreements with the Mediterranean countries, where a major conflict between the CAP and the Association Agreements is developing, as discussed in Chapter 15). A much more important matter for many LDCs has been the way the CAP cereals policy depresses and destabilises world cereal prices and it is on this issue that the present chapter concentrates.

THE CAP AND THE WORLD MARKET

The current Uruguay round of international trade negotiations under the GATT aims to liberalise trade in agriculture. Domestic support policies need to be made more responsive to international market signals, thereby reducing the trade-distorting influences that inevitably follow when the US and the EC, in particular, compete with each other to subsidise exports of surplus production. The Punta del Este Declaration at the outset of the Uruguay round recognised the need to accord special and differential treatment to developing countries, and ways are being sought to take into account the possible negative effects on net food-importing LDCs of agricultural trade reform .

In order to maintain prices paid to European farmers above levels determined in world markets, the CAP uses a system of variable import levies and variable export subsidies, combined with the provision for intervention buying, as described in Chapters 1 and 12. The CAP effectively insulates the EC domestic agricultural sector from world market forces and trends, in order to maintain internal Community objectives of farm income support, stable prices and secure food supplies. The consequences of this policy are well known - a rapid growth in production, increased self-sufficiency ratios, surpluses, income transfers from consumers and taxpayers to producers, financial transfers between member states and a continuous budgetary burden. Of particular note in the United Kingdom has been the doubling of wheat production since 1971 (the UK became a net cereal exporter in 1981) and the substantial increases in self-sufficiency in dairy products and sugar, which have reduced dependence upon imports (Chapter 4). Caribbean countries have seen their traditional export markets for sugar reduced as domestically produced sugar-beet is substituted for imported cane sugar within Europe. Botswana's beef exports have been similarly hit. The European Community has emerged as a major exporter of agricultural produce in recent years and is now the second largest agricultural exporter after the United States. European Community incursions into export markets, plus a reduced EC import market for traditional suppliers, including LDCs, are a growing source of tension in the international trade arena.

The question 'what would be the situation if European Community policies were wholly or partly removed?' is a complex one already confronted in Chapters 7 and 8. To answer it one must first assume alternative EC arrangements to protect domestic agriculture, and then model the new world trade flows that would ensue. Europe is so large that any change in its level and methods of protection will have complex repercussions on the whole of the world trading system. Cross-linkages between world agricultural prices are diverse and the substitution mechanisms of demand and supply are not obvious.

Exercises of this type calculate the associated changes in trade balance and welfare for a variety of trade liberalisation scenarios. Burniaux and Waelbroeck (1988) use a general equilibrium model to calculate that, with no agricultural protection in Europe (the free trade option), output would fall by 16.8 per cent and world agricultural prices would rise substantially, with grain prices increasing by 13.4 per cent. Colman (1984) summarises some results and suggests that the overall benefits to the LDCs as a whole, from the reduction of agricultural trade barriers, would be quite small. The CAP exercises a major constraint upon the agricultural export earnings of less developed countries, but offsetting this many of the poorest LDCs, which are becoming increasingly dependent upon food imports, would be adversely affected by any policy changes which caused international food prices to rise. However, higher world food prices should force LDC governments to pay more attention to domestic agricultural production and rural development.

It is generally accepted that, in addition to the effects noted above, agricultural policies, such as the CAP, force price instability on to world markets to the detriment of less developed countries. It is held that stabilising Community farm prices has increased price instability on world markets. The issue of instability and variability of world markets is an important one, and for illustrative purposes the following two sections examine, in some detail, the world market for cereals.

THE WORLD CEREALS MARKET

Recent trends in the world cereal market are shown in Table 13.1 for wheat and coarse grains (little rice is traded on the world market). Despite the headline-making problems in starving Africa and droughts in the United States, we are entering an era of grain surpluses, buyers' markets and high storage costs of surplus grain stocks. Production has expanded by 40 per cent since 1974, more than matching effective world consumption. A growing number of exporters face a shrinking set of importers and the long-term trend of prices is downwards, with the weakness of the US dollar not providing much stimulus to demand.

There is intense competition by exporters to dispose of surplus stocks in world markets, involving a variety of concessional credit programmes and food aid, but the capacity of some deficit areas to import is constrained by lack of foreign exchange. As ever, the issue of food aid is clouded with political and logistical questions. In the future, world production is expected to continue to grow and export competition to increase. Surpluses will increase and prices will tend to move downwards. Lack of income growth in less developed countries will curtail consumption increases for grain, both for direct human usage and as livestock feed. With the costly process of stockholding there may be increased pressure to apply production controls (quotas or set-aside schemes) in developed countries, but income support to farmers is likely, as in the past, to override both cost and trade considerations.

Table 13.1: World Wheat and Coarse Grains Trade, 1974/75-1988/89 (totals in million tonnes)

	1974/75	1979/80	1988/89*
Total World Production	976.8	1,164.7	1,378.0
United States (as % of total)	20.4	25.5	16.2
Other major exporters**	9.8	7.9	7.4
Western Europe	14.5	12.6	15.7
Soviet Union	18.8	14.7	14.8
Eastern Europe	9.3	7.8	7.3
People's Republic of China	10.0	12.5	14.7
Others	17.1	18.9	23.9
Stocks	115.1	172.0	277.8
United States (as % of total)	23.5	44.9	n.a.
Other	76.5	55.2	n.a.
Total Trade	127.6	186.8	224.0
Exports:			
United States (as % of total)	48.9	58.2	50.9
Other major exporters**	34.4	29.7	20.9
Western Europe	10.0	8.9	18.0
Soviet Union	3.9	0.3	0.3
Others	3.4	2.8	9.9
Imports:			
Western Europe (as % of total)	25.7	16.4	4.3
Soviet Union	4.1	16.3	20.3
Japan	14.5	13.1	13.8
Eastern Europe	8.7	9.4	4.2
People's Republic of China		5.8	8.4
Others	46.9	39.0	49.0

* Estimate.
** Other major exporters are Argentina, Australia, Canada, South Africa and Thailand.
Source: US Department of Agriculture and International Wheat Council.

Cereal supply is crucial to solving the malnutrition problems in less developed countries. Green and Kirkpatrick (1980) estimated that cereals contributed 65 per cent of the total calorie supply in the 'most seriously affected' malnourished countries in the 1970s. FAO (1983) suggests that up to 165 million people in Africa will be at risk by the year 2000 if grain availability does not improve, and increasingly we see less developed countries becoming dependent on cereal imports. Huddleston (1984) estimates that less developed countries' cereal imports, largely of wheat, have risen from 30 million tonnes to 97 million tonnes between 1960 and 1981. In general, agricultural production growth in LDCs is failing to keep pace with growth in demand arising out of increases in population and income. This is especially the case in Africa.

The developing countries' share of the world market in foodstuffs, which is predominantly cereals, was estimated to be 45 per cent in 1981 and imports and food aid continue to rise. The USDA (1986) reported that imports of grain into sub-Saharan agriculture were over 12 million tonnes in 1985, only 10 per cent below the previous year's record level. Wheat accounted for 50 per cent of the total, rice and corn for 22 per cent and 15 per cent respectively. The European Community has emerged as the major agricultural supplier, supplying the region with one-third of its import needs since 1980, whilst the United States has held only a 15 per cent share. Unpublished work at Newcastle University shows that per capita food production for 16 sample countries in Africa declined by 32 per cent between 1964 and 1982, whilst at the same time imports rose by 128 per cent. The greater part of this rise occurred after 1972. In the foreseeable future there will be a continuation of the trend for LDCs to increase their imports of temperate zone food, implying that international trade will remain an important instrument as these countries attempt to stabilise domestic consumption levels.

Of increasing concern is the increased variability in world cereal production and prices in recent years. The causes of this phenomenon are only partially understood, but are linked to individual crop yield variability associated with the adoption of new seed varieties, fertiliser-intensive technologies, a reduction in offsetting patterns between different crops and regions and periodic oil price shocks. Hazell (1985) calculates that world cereal production grew at 2.7 per cent per annum between 1960/61 and 1982/83 but that there was a widening band of variability around the trend underlying this increase. Total cereal production rose by 305 million tonnes, of which wheat and maize each contributed 33 per cent of the increase, rice 12 per cent and barley 18 per cent. Comparing the two periods 1960/61 to 1970/71 and 1971/72 to 1982/83 the variability of production increased markedly in the second period particularly in the wheat/barley regions of Africa, South America, Oceania, Canada and the USSR. Rice regions are less variable since much of the crop is irrigated. In terms of African food security, the probability of a 5 per cent shortfall below the trend of production is 40 per cent in wheat/barley regions, 18 per cent for sorghum/millet, 26 per cent for maize and only 11 per cent for rice.

It is not easy to account for the increased production variability. Similar technologies, similar weather, reduced offsetting yield effects between different cereals, oil price effects, the expansion of crops to marginal areas, fewer drought-resistant crops and a narrowing genetic base all play their part along with the pricing policy for these commodities. However, a continued high level of variability in world cereal production and hence prices seems likely. Inevitably, stockpiling to cope with this will be a costly operation, and the developed countries may be unwilling to hold and fund world grain stocks in the future just to ensure food security for the less developed countries.

INSTABILITY IN THE WORLD WHEAT MARKET

Taking the fob price of US exports of wheat as indicative of price on the world market, Figure 13.1 shows the world price of wheat over the quarter century to 1986. In nominal dollar terms (that is, unadjusted for inflation) the world price showed remarkable uniformity from 1960 to 1971, at around US$60 per tonne. The commodity crisis of the early 1970s resulted in a tripling of this price by 1973. Since then it has undergone considerable fluctuation, and in 1986 stood at around US$110 per tonne.

Figure 13.1:   World Price of Wheat (Nominal Prices)(US$/tonne)

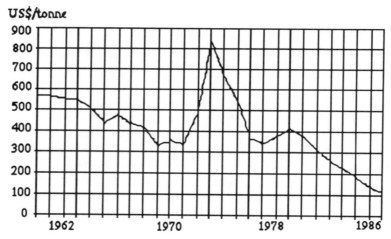

Source: International Wheat Council.

In Figure 13.2 these nominal world market prices have been deflated, using a world GDP index, to show prices in 1986 values. In real terms the world market price had been falling steadily prior to the commodity crisis and, indeed, has been falling since. If the abnormal prices of the mid-1970s are discounted, world market price exhibits a continuing downward trend. Wheat prices have fallen substantially in real terms.

The mean price, expressed in 1986 values, over 1960-86 is US$415 per tonne, but this average conceals price variability during the period. As a measure of the extent of this variability it is useful to calculate the coefficient of variation (c.o.v.). This statistic relates the standard deviation to the mean, and is expressed on a percentage basis. The higher the c.o.v. the greater the degree of fluctuation of price about its mean level. The c.o.v. associated with the mean of US$415 tonne is 38.2 per cent.

Figure 13.2: World Price of Wheat (1986 Prices) (US$/tonne)

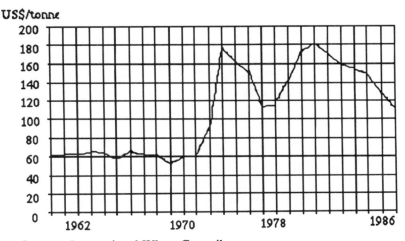

US$/tonne

Source: International Wheat Council.

In an attempt to isolate the effects of the commodity price boom of 1972-75 and to examine more closely what has been happening to world prices over time, let us divide the 27-year period into three separate periods, representing the pre-commodity crisis period (1960-71), the commodity crisis period itself (1972-75) and the post-commodity crisis period (1976-86). The mean and c.o.v. for each of these periods are shown in Table 13.2. The three periods exhibit markedly different summary statistics. For reasons that are well documented, world price increased dramatically in the early 1970s, but perhaps of greater surprise is the equally dramatic fall in price in the last of the three periods. At US$285/tonne the average price of wheat since 1976 is only 62 per cent of the average price during the 1960s. The coefficient of variation rises from 18.5 per cent through 21.7 per cent to 35.2 per cent over the three periods, suggesting increased price variability. On examination of Figure 13.2 it can be seen in both the first and third periods that the downward trend in price is quite smooth. Over the period 1960-71 the world price of wheat fell at an average rate of 5 per cent a year; since 1976 this annual average rate of decline has risen to 11 per cent.

Table 13.2: World Price of Wheat

Period	Mean (US$/tonne; 1986 prices)	c.o.v. (%)*
1960-71	462	18.5
1972-75	635	21.7
1976-86	285	35.2
1960-86	415	38.2

Source: Compiled from International Wheat Council data.
* The coefficient of variation (c.o.v.) is calculated as the standard deviation as a percentage of the mean.

Let us now turn to look at the EC's external trade in wheat. Figure 13.3 shows the EC's exports and imports (intra-EC trade is excluded) of wheat since the establishment of the common market for cereals in the late 1960s. Exports of wheat and wheat flour have increased from a low of 3.1 million tonnes in 1970 to a high of 17.3 million tonnes in 1984, whilst imports have fallen from a high of 7 million tonnes in 1972 to a low of 2 million tonnes in 1984. The difference between the two sets of bar charts in Figure 13.3 represents the Community's net external trade in wheat and wheat flour. The Community was a net importer in 1970, 1972, 1976 and 1977, but in more recent times exports have substantially exceeded imports, with net exports peaking at 15.2 million tonnes in 1984.

Figure 13.3: EC Exports and Imports of Wheat (Including Wheat Flour)

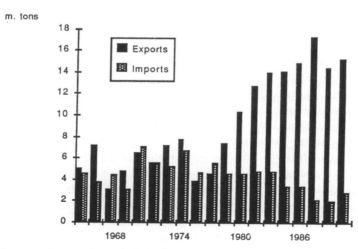

Source: International Wheat Council.

Table 13.3 shows that in the period 1968-75 exports of wheat from the Community averaged 5.9 million tonnes with a c.o.v. of 25 per cent. Net exports over the period averaged only 0.9 million tonnes per annum, but were subject to significantly greater variability, as is indicated by the c.o.v. of 167 per cent. In the post-commodity crisis period, whilst gross exports doubled, net exports increased by more than eightfold. Again, variability of net exports during this period was high, though less than over the previous period.

Table 13.3: World Price of Wheat

Period	Gross Exports		Net Exports	
	Mean (m. tonnes)	c.o.v. (%)	Mean (m. tonnes)	c.o.v. (%)
1968-75	5.9	24.9	0.9	167.2
1976-86	11.7	37.0	7.8	66.7
1968-86	9.2	48.5	4.9	109.2

Source: Compiled from International Wheat Council data.

Thus, a cursory examination indicates that, over the last thirty years, the real price of wheat on the world market has been on a downward trend and has exhibited somewhat greater instability, whilst over the last twenty years net exports from the EC have risen dramatically. However, attributing any cause and effect between these occurrences requires a fuller analysis. United States exports have risen dramatically as well, and they still account for a large share of the world market. Other developed country exporters have also contributed to the destabilisation of world markets, and it is difficult to isolate EC influences alone.

The impact on the world price of EC exports of wheat is generally regarded as being twofold. First, any increase in the Community's excess supply (net exports) can be expected to depress the world price. Second, as a consequence of the workings of the cereals regime within the CAP, variability of world price can be expected to increase. There have been a number of studies illustrating these effects. For example, Sarris and Freebairn (1983) use their own model of the world wheat market to show that if the EC were to dismantle the CAP and adopt the unlikely position of free trade, the world wheat price would increase by 9.2 per cent (an example of the price-depressing effect of the CAP). Moreover, the c.o.v. of world price would decrease by 19.8 per cent (an example of the instability-increasing effect of the CAP). If the entire world followed suit and moved to free trade, Sarris and Freebairn estimate the world price would rise by 11 per cent and the c.o.v. fall by 35.3 per cent (an illustration of the effects on the world market of other countries' agricultural policies). Using these results, Sarris and Freebairn conclude that the CAP is responsible for 84 per cent of the price depression in the world market and for 56 per cent of increased price variability, as compared with a position of global free trade. Comparable results to those of Sarris and Freebairn have been obtained from other studies.

The point has already been made that the EC is not alone in protecting its agricultural sector. Most developed countries throughout the world subsidise their domestic agricultural production via a variety of policy instruments and measures. In an attempt to compare levels of protection around the world, the concept of 'producer subsidy equivalent' (PSE) has resurfaced. This seeks to convert to a common base all forms of support by showing 'the payment which would be required to compensate farmers for the loss of income from the removal of a given policy measure' (OECD 1987, p 100). In Figure 13.4 PSEs are shown, as a percentage of production values, for wheat and rice, for the EC and a selection of countries. The general pattern exhibited by these data is now fairly well known. Agricultural protection is high in the EC, the US and Japan, much lower in countries like Australia and New Zealand, and negative (that is, serving as a production tax) in some developing countries.

Figure 13.4: Wheat and Rice PSEs, 1982-86 (as % of Producer Receipts)

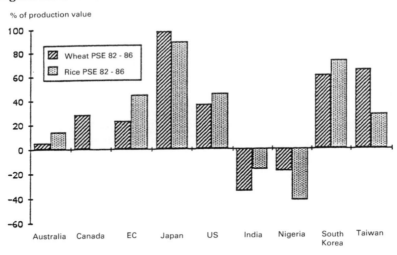

Source: Harvey (1988) after USDA (1987).

Within the developed world the level of protection is by no means the same for all agricultural commodities. Dairy products and sugar tend to receive the highest levels of subsidy, with the grain-based 'white' meats typically at the other end of the scale, this being partly a result of the protection given to cereals. An indication of the pattern of protection by commodity in the European Community, the United States and Japan is given in Figure 13.5.

Figure 13.5: PSEs by Commodity, 1982-86

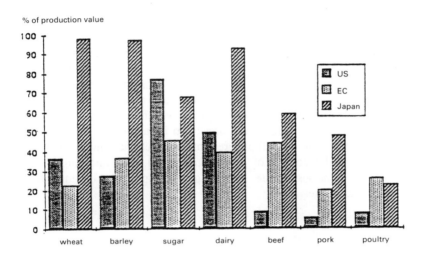

Source: Harvey (1988) after USDA (1987).

Figure 13.6: Level and Stability of World Price and Trade Volume Changes from Industrial Countries Liberalisation

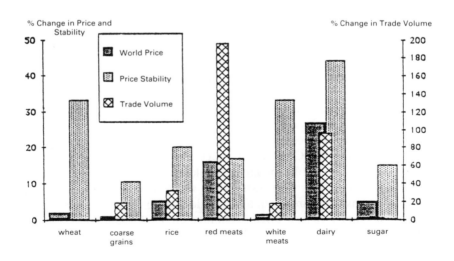

Source: Harvey (1988) after Tyres and Anderson (1986).

251

Reference has been made already to the effect of the CAP on the level and stability of world prices. Some authors have attempted to show the impact on world markets of the elimination of all forms of agricultural support in the developed world. The work of Sarris and Freebairn (1983) was cited earlier; other examples include Tyres and Anderson (1986) and the OECD (1987). Whilst the scope of these studies is beyond the remit of this chapter, the results do highlight the problems faced by LDCs as a direct consequence of agricultural protection in developed countries. As an example, consider Figure 13.6. This shows the impact that such liberalisation would have on the level and stability of world prices, and the changes in trade volume, for a selection of commodities. Although there would appear to be a good deal of variation between commodities, the overall picture remains much the same. Agricultural protection reduces the volume of world agricultural trade and causes world prices to be lower and less stable. Of this global impact, the EC must bear a major share of the blame.

EFFECTS OF INSTABILITY ON LESS DEVELOPED COUNTRIES

Fluctuations in production levels and hence food availability frequently occur in LDCs consequent upon uncertain and variable climates and production environments, coupled with a less sophisticated traditional infrastructure. Food security (the ability of LDCs to meet target levels of consumption on a year-to-year basis) is not easy to plan. Access to assured imports at known prices would help the planning process. Reliance on imports by food-deficit LDCs requires that they have confidence that grain imports would be available in the future within some foreseen price range. In addition they must have access to the foreign exchange needed to maintain consumption given shortfalls in domestic production. The combination of domestic production instability and world price instability makes effective planning well-nigh impossible or infeasibly expensive. In years of world shortages the rich food-importing nations, like the Soviet Union and China, can always outbid the poor.

It is now largely accepted that the CAP has achieved internal price stabilisation within the EC but at the expense of increasing instability outside the Community. The burden of adjustment to changes in global production of agricultural commodities is thrown on the world market, resulting in larger price fluctuations, compared with a free trade situation. Liberalising the CAP would substantially reduce price instability on world commodity markets. This argument also extends to export markets of less developed countries. Increased price and revenue instability, brought about by the CAP, adds to the production risks of agricultural exporters of the Third World, undermining investment and future productivity (Roarty 1985).

Production and price fluctuations in wheat in developed countries can be a potential problem for the rest of the world. Some changes in production directly reflect protectionist policy decisions of the United States, the European Community and Japan. Greater stability in production levels would be to the advantage of developing countries in planning their food supplies and ensuring food security. Reducing some variations in production may require developments in technology and husbandry, but eliminating policy-induced variations is an urgent priority for the international community. The CAP is therefore an adverse influence on the food security of less developed countries. Any improvement must involve policy within the EC, adjusting so that stocks and prices are more sensitive and responsive to world market conditions.

TRADE PREFERENCES

The European Community operates a number of preferential trading agreements with developing countries. These offer certain trade advantages over other third countries by making access to the Community market somewhat easier. Of the agreements, the most elaborate is the package of arrangements under the Lomé Convention, covering nearly seventy African, Caribbean and Pacific (ACP) countries. The first Lomé Convention became operational in 1975, since when there have been two updates, with the period 1985-90 now covered by Lomé III. Under Lomé, many exports from ACP countries enter the Community free of customs duties and free of quantitative restrictions, whilst the ACP countries themselves are not required to offer reciprocal trade concessions. The European Commission sees the Lomé Convention as 'a symbol for the Community and of North-South relations - and therein lies both its strength and its weakness, for the hopes it has generated turn to disappointment at its slightest shortcoming' (European Commission 1985). As this statement implies, Lomé is not without its critics.

The Lomé Conventions are quite separate from the CAP, though some common ground does exist. The principle of free, or freer, access to the Community market does not extend, in general, to those commodities covered by the CAP. There are, however, special trade arrangements for sugar. Sugar is a commodity which can be produced in both temperate and tropical regions, and is an important crop in both the Community and some ACP countries. Under the sugar protocol, ACP countries [1] not only have free access to the Community market for around two-thirds of their sugar exports, but also benefit from the price they receive being tied to the high support price paid to

1   The ACP countries in question are: Mauritius, Fiji, Guyana, Jamaica, Swaziland, Barbados, Trinidad and Tobago, Belize, Zimbabwe, Malawi, St Christopher and Nevis, Madagascar, Tanzania, Congo, Ivory Coast, Kenya, Uganda and Surinam.

European farmers. This arrangement has proved to be fairly controversial, as indeed has the sugar regime under the CAP. A ceiling of 1.3 million tonnes of ACP sugar is covered by the arrangement, which means, in effect, that surplus sugar production within the Community (which is around 2.5 million tonnes) is increased by an equivalent amount. This excess is disposed of on the world market, depressing world price and thereby lessening the export earnings of those developing countries not party to the protocol.

One special feature of Lomé is a system of 'compensatory finance' known as Stabex. This is a form of insurance that operates in bad harvest years and seeks to compensate ACP counties for losses in export earnings, thereby attempting to stabilise their export income. Fifty commodities are covered by Stabex, [2] but, whilst it may go some way to providing temporary relief in times of crisis, it operates within a limited budget and cannot preclude serious fluctuations in export earnings. Once again, the commodity coverage leaves the CAP largely unfettered.

REFORM OF THE COMMON AGRICULTURAL POLICY

For many years now reform of the Common Agricultural Policy has been a popular topic for debate (see Chapter 16). The CAP is frequently criticised both from within the Community and from interested parties in third countries. The latter include not only the LDCs but also the world's other main agricultural traders, in particular the United States, Canada, Australia and New Zealand. Indeed, the United States has threatened agricultural trade wars. In response to these criticisms there have been minor changes in the workings of the CAP over the years, but the principle of high domestic support prices remains unchanged. Critics of the CAP, however, are constantly hoping for major reform.

The criticism from within the Community stems mainly from the financial cost of the CAP. Budgetary expenditure (and to a lesser extent the consumer cost) has been the major concern of those seeking reform. The budgetary cost of the CAP is financed principally by the Community's taxpayers, and as this cost has escalated as a consequence of increasing surpluses, so the demands for reform have grown. Consumers in the Community, as we saw in Chapter 7, also pay heavily for the CAP, but as yet have consistently failed collectively to make their presence felt. Since the burden on consumers does not increase as production increases, there is little reason to suppose their pressure will be any more effective in the future.

However, there are signs that environmental pressures against intensive agriculture are growing particularly from the 'Green Movement' within European politics. Major reform, if and when it occurs, is likely to be in response to these internal pressures. If policy-makers are reluctant to heed the

2   The main commodities are coffee, cocoa, cotton, groundnuts, sisal, timber, oilcake, bananas, tea and palm products.

criticisms of their own taxpayers, it is extremely unlikely that they are going to be swayed by the claims of third countries. Thus, pressure from the LDCs cannot be expected to have any real impact in reforming the CAP. Probably the best that third countries can hope for is that politicians in the Community respond to the growing air of disquiet from their own electorate. In any case, many LDCs must also put their 'own house in order' with respect to the pricing policy for agricultural output and inputs. Food prices play a dual and conflicting role in the elimination of poverty and hunger; on the one hand they provide an incentive to production, thus generating employment and income; on the other they help determine the entitlement to food, the extent of hunger and malnutrition and thus real income levels. Short-sighted governments in some less developed countries often keep the price of foodstuffs low for the urban proletariat, thereby undermining incentives to local farmers and building up a dependence on imported foods. Policy changes are required in many less developed countries to overcome the past neglect of agriculture.

International political pressure should be directed to convincing the EC that it is responsible for some of the instability in world markets. Accepting its responsibilities, it may be possible to institute a continuing food aid programme whereby aid donors provide food aid to meet grain production shortfalls which exceed, say, 6 per cent of the trend level of production. In 1986, 7.2 million tonnes of cereals and 121,400 tonnes of milk products were supplied to less developed countries as food aid by the EC. The total value of this aid, taking account of transport costs and the differences between European and world prices, is estimated at 412 million ECU. Matthews (1985) suggested that another route to reform would be to maintain the average level of protection to EC agriculture by means of a constant tariff, so that fluctuations in world prices were reflected in EC markets. Permitting the EC market to react to world price fluctuations would help dampen those fluctuations which occur. This would be a positive benefit to less developed countries by lowering the probability that they could not afford food imports in periods of high food prices. However, it would mean that EC farmers could no longer depend on a guaranteed minimum price for their produce, while EC consumers would find food prices more variable. There is thus a direct trade-off between the food security of LDC populations and price stability within the EC.

In terms of potential changes to the CAP, major reform is likely to involve either significant price cuts for the main agricultural commodities, including cereals, or physical controls on production. Either would result in a reduction in the surpluses and thus in EC exports. However, from the experience of recent years, reform is far from a foregone conclusion. The CAP has a history of lurching from crisis to crisis. Continued criticism from the less developed countries and other third countries will add to the ground swell of pressure for reform, and may hasten the process of change, but should not be expected to achieve any great degree of success on its own. Firm conclusions about reforming the CAP from an LDC perspective are difficult. LDCs are heterogeneous, and higher world food prices, as explained earlier, would have different consequences for different groups both between and within LDCs.

Nevertheless, internal pressures are likely to prevent any worsening of the impact of the CAP on world markets in the longer term. Budgetary pressures alone should be sufficient to limit the growth of EC exports. However, a continued internal EC emphasis on coping with crises in a partial, commodity-by-commodity, approach may well mean that international markets are subjected to increasing short-term fluctuations and uncertainties, as evidenced by the consequences of milk quotas on beef markets. Unfortunately, living through the 'short term' is the overriding ambition of many LDCs, and even of developed countries reliant on agricultural exports. Given the time taken by the EC to reach decisions about inevitable policy changes, the short-term uncertainty on world food markets is likely to continue.

REFERENCES

Burniaux, J.M. and Waelbroeck, J. (1988) Agricultural Protection in Europe. In: Langhammer, R.J. and Rieger, H.C. (eds) *ASEAN and the EC: Trade in Tropical Agricultural Products,* ASEAN Economic Research Unit, Singapore.

Colman, D. (1984) EEC Agriculture in Conflict with Trade and Development, *Manchester Papers on Development,* No. 10, pp 1-12.

European Commission (1985) *Europe-South Dialogue,* Directorate-General for Information.

FAO (1983) *The State of Food and Agriculture - The Situation in Sub-Saharan Africa,* FAO, Rome.

Green, C. and Kirkpatrick, C. (1980) *A Cross Section Analysis of Food Insecurity in Developing Countries - Its Magnitude and Sources,* Discussion Paper No. 1807, Department of Economics, University of Manchester.

Harvey, D.R. (1988) *Food Mountains and Famines: the Economics of Agricultural Policies,* Discussion Paper 5/88, Department of Agricultural Economics and Food Marketing, University of Newcastle upon Tyne.

Hazell, P.B.R. (1985) Sources of Increased Variability in World Cereal Production Since the 1960s, *Journal of Agricultural Economics,* 36(2), pp 145-159.

Huddleston, B. (1984) *Closing the Cereals Gap with Trade and Food Aid,* IFPRI Research Report No 43, IFPRI, Washington DC.

International Wheat Council (various years) *World Wheat Statistics,* Annual Publications, International Wheat Council, London.

Matthews, A. (1985) *The Common Agricultural Policy and the Less Developed Countries,* Gill and Macmillan, Dublin.

OECD (1987) *National Policies and Agricultural Trade,* Paris.

Roarty, M.J. (1985) The EEC Common Agricultural Policy and its Effects on Less Developed Countries, *National Westminster Bank Quarterly Review*, pp 2-17.

Sarris, A.H., and Freebairn, J. (1983) Endogenous Price Policies and International Wheat Prices, *American Journal of Agricultural Economics*, 65, pp 214-224.

Tyres, R. and Anderson, K. (1986) *Distortions in World Food Markets: a Quantitative Assessment,* Background Paper for the World Bank's World Development Report, Oxford University Press, New York.

USDA (1986) *Sub-Saharan Africa - Situation and Outlook Report,* Economic Research Report, Washington.

USDA (1987) *Preliminary Estimates of PSEs and CSEs, 1982-1986,* Economic Research Service, Washington.

CHAPTER 14

THE CAP AND THE UNITED STATES

Tim Josling

INTRODUCTION

At the end of May 1989, the United States (US) named selected trade practices of three countries - Japan, Brazil and India - as special targets for negotiation and potential retaliation under the so-called 'Super 301' provision of the 1988 Omnibus Trade Bill. The indictments specified trade barriers in the areas of telecommunications, super computers, investment services and intellectual property rights. From the rhetoric across the Atlantic in recent years, one might have expected to see the European Community (EC) named as a Super 301 country, for its grain levies, beef hormone ban and oilseed subsidies.

That such a nomination was not made reflects the complexity of US-EC relationships. Issues of agricultural trade are the focus of multilateral negotiations going on in Geneva. Agreement was reached in April 1989 on the agenda for the next two years of the negotiations. It was presumably felt that such an action singling out the EC would not have been constructive at the present time. Moreover, commercial relationships between the US and the EC have improved with the Bush administration, and co-operation rather than confrontation seems to be on the rise. But tensions over agriculture remain only just beneath the surface, and could erupt again at any time.

The cause of these tensions owes much to the emergence of EC agriculture as a major exporter in temperate zone markets. The US has seen its exports to the EC diminish over the last decade, and has faced increasing competition in third country markets. EC exports have also penetrated the US domestic market to an increasing extent. The US has a strong belief in the capacity of its own agriculture to compete on level terms with other countries. Its objection is to subsidised competition from other countries which seek to use world markets to avoid domestic resource adjustments and hence prevent the true competitiveness from emerging. The European Community is singled out as a major player in the game of subsidised exports and high domestic market prices. This chapter reviews some of these notions - not so much with the idea of seeing who is 'in the right', but of helping to understand the differing viewpoints.

UNITED STATE-EUROPEAN COMMUNITY TRADE DEVELOPMENTS

Trade tensions broadly reflect the severity of domestic policy problems and the state of world markets. The exacerbation of tensions since 1980 is a clear reflection of these factors. The growth of EC agricultural exports is documented in other chapters in this book. Figure 14.1 shows a comparison of EC and US agricultural exports for the period 1970 to 1987. During the 1970s, US exports rose rapidly. The US had in place the capacity to meet the surge in demand from the USSR and from the developing world. European exports lagged until 1977, but accelerated with the generally favourable trade conditions until 1981. That year represented a peak for US agricultural exports, at $43 billion. Over the next few years, US exports declined dramatically, to a low point of $26.1 billion in 1986. EC exports stalled a little then recovered, and in 1986 were actually higher than those of the US. The European Community, always thought of as the 'largest import market' for US agricultural products, was now the world's largest exporter. A recovery for the United States since 1987 has helped to restore somewhat the earlier balance.

The trend in overall exports sets the tone for United States farm policy. The budget cost in essence reflects the balance between domestic production and export demand. Escalation of budget costs over the 1980s was a major policy concern in the United States as it was in the European Community. Figure 14.2 shows the outlay on price supports over this period in both the EC and the US. In the 1970s, farm support costs stayed low, with buoyant export markets and firm prices. Support costs in the US took off in the wake of the drop in exports in 1982, coupled with the generous policy prices set under the 1981 Farm Bill. A drought-induced drop in production in 1983, coupled with the cost-shifting effect of the payment-in-kind (PIK) scheme, gave a temporary respite, but, by 1986, expenditure had reached a peak of $25 billion. Indeed, as European Community representatives in Brussels and Washington were quick to point out, US support costs exceeded those in the EC at that time. More recent developments have also returned this relationship to its more normal state.

Bilateral trade figures point out the same story. Table 14.1 shows bilateral trade between the United States and the European Community over the period 1975-86. Over the last half of the 1970s, trade was increasing steadily in both directions. The situation changed abruptly after 1980. United States exports of agricultural products to the European Community of 10 member states fell from a high of $9.6 billion in 1980 to $5.2 billion in 1985, recovering the next year to $6.6 billion, in part because of the enlargement of the Community in 1986. By contrast, EC agricultural exports to the US continued to rise, reaching $4.1 billion in 1986 ($3.8 billion for the European Community of 10). That same year, the European Community of 12 sold $71 billion of non-agricultural goods to the United States and purchased $44 billion of those

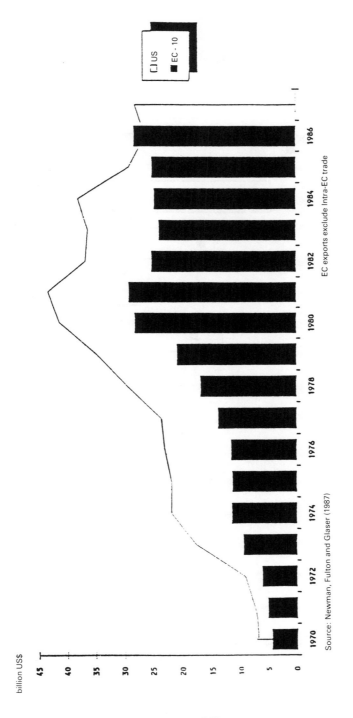

Figure 14.1: Agricultural Exports, US and EC, 1970-1987

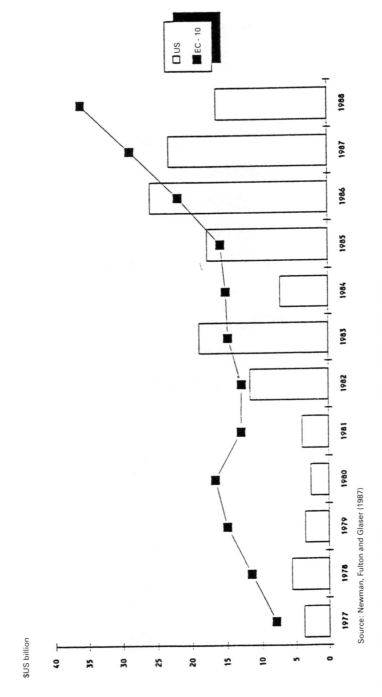

Source: Newman, Fulton and Glaser (1987)

Figure 14.2: Outlays for Price and Income Support EC and US, 1977-1978

goods. The agricultural trade surplus of $2.5 billion went only a small way to offsetting the US merchandise trade deficit with the Community. [1]

Table 14.1: United States - European Community Bilateral Agricultural Trade

	(billion dollars)	
Year	US to EC	EC to US
1975	6.0	1.1
1976	6.8	1.3
1977	6.9	1.4
1978	7.5	1.9
1979	8.1	1.9
1980	9.6	2.1
1981	9.1	2.3
1982	8.4	2.5
1983	7.4	2.8
1984	6.5	3.2
1985	5.2	3.6
1986	6.6	4.1

Source: Newman *et al.* (1987).
Note: EC-12 figures are given for 1986:
EC-10 trade flows are 5.6 and 3.8 billion,
respectively, for that year.

The fall in US exports to the EC in the 1980s was primarily in the 'big-ticket' items of grains, non-grain feed ingredients and oilseeds and products (see Table 14.2). Corn sales were hard hit by emerging surpluses of barley in the EC and by the continued attractiveness of non-grain feed ingredients in livestock rations. Soybean sales were hit by increased oilseed production in the EC, which grew from 3.2 million tonnes in 1980/81 to 11.6 million tonnes in 1987/88. Domestic oilseed production now accounts for nearly one-half of the total EC oilseed use (USDA 1988). More recently, other markets have been affected. Beef and variety meat sales to the EC from the US had reached over $100 million in 1988, when the ban on imported meat produced with hormones was introduced. Cotton sales fell away after 1985, in part as a result of Spanish and Portuguese accession, and the US has lost some of its small share in the EC wine market in recent years.

---

1   Economists may point out the irrelevance of bilateral deficits in a multilateral trade system. Such balances do, however, play a significant role in shaping trade policy attitudes.

Table 14.2: United States Agricultural Exports to the European Community, Selected Products ($m). Fiscal Year, Exports to EC(12)

Commodity	1982	1983	1984	1985	1986
Oilseeds & Products	5,173	4,403	3,378	2,318	2,506
Grains & Feeds	3,403	2,488	2,621	1,800	1,507
Animals & Products	987	788	793	649	765
Tobacco Manufacturing	616	636	669	663	549
Nuts & Preparations	301	250	263	330	257
Fruits & Preparations	229	183	156	136	161
Cotton	215	209	369	375	123
Vegetables & Preparations	178	152	147	128	137

Source: Newman *et al.* (1987).

TWO AMERICAN VIEWS OF THE CAP

To an economist, such bilateral trends are the result of a complex of factors in world markets and in the domestic economies of the trading partners. Some of these factors relate to yield trends and fluctuations and to technical changes in sectors that use agricultural raw materials. Other factors are imposed on agriculture by macroeconomic developments, such as exchange rate changes. Others relate to world market changes and the behaviour of other trading countries. And some part of the trend is undoubtedly due to policies implemented in the importing country.

This somewhat academic (in the sense of detached, rather than irrelevant) view can be found in the US. Analysts in the Economic Research Service (ERS) of the United States Department of Agriculture have been in the forefront of a wave of model building which has sought to relate macroeconomic and agricultural policies to trade developments. More recently, there has been considerable experimentation with economy-wide models which elucidate the impact of agricultural policy on non-farm sectors. The result of such studies gives us a much clearer picture of the implications of government actions in agriculture, and allows one to view changes in bilateral or sectoral trade in a wider perspective. For the sake of easy reference, this can be called the 'ERS view'. [2]

To those concerned with export policy, such nice distinctions are less appealing. If a lucrative market is shrinking, then someone must be responsible. Governments set policies and policies influence trade. Foreign supplies generally receive less favourable treatment in domestic markets, and

---

2   Such a label does not imply that all individuals within the organisation hold these views, or that this is an official view of the agency.

governments are continually finding ways of hindering imports of agricultural products. When imports drop, the first place to look is government policy. In the case of trends in US-EC trade, so the argument goes, the culprit is obvious. The CAP is clearly to blame.

This tendency to hold governments accountable for trends in bilateral trade balances is not confined to agriculture, nor is the European Community alone in being under pressure to change policies for these reasons. But the way that this view comes to dominate commercial relationships is perhaps reflected as clearly as anywhere in the United States' attitude towards the CAP. It might be called the 'FAS view', after the Foreign Agricultural Service of the USDA. [3] The FAS view is widely shared on Capitol Hill and among the agricultural exporting interests: it has shaped US foreign trade policy in agriculture for the past decade.

The FAS view explains much of the rhetoric that has shaped transatlantic discussion of agriculture. The list of trade conflicts, from the 'Chicken War' of the 1960s, to the beef hormone dispute of 1989, is a testament to the view that confrontation is the only way to get Europe's attention and halt protectionism. The result of this approach has been some modest success in preventing new trade barriers, and a considerable legacy of anti-US feeling among EC policy-makers. Whether the broad development of the CAP has been influenced is less clear.

CURRENT TRADE FRICTIONS

To illustrate this attitude it may be worth looking at some recent US-EC agricultural trade irritants. At the risk of caricature, the predominant US view of these events will be stated in fairly succinct terms. In fact, the direct, straight-from-the-shoulder approach to such matters is itself a feature of the predominant view. Equivocation and qualification are clearly an attempt to prevaricate and obfuscate the situation.

Soybean Sales and Feed Ingredients

The longest-running battle is over soybean exports to the European Community. The United States received a concession from the Community in the Dillon round of the GATT that the tariff on soybeans and meal would be bound at zero. [4] This one action has turned out to have dominated, more than any other, the next 25 years of trade relations. To the EC it was a concession easily made, since trade was fairly small in those commodities and domestic production almost non-existent. Low tariffs on imports of oilseeds

---

3   The same caveat applies as in the previous footnote.

4   The concession was actually negotiated in the Article XXIV (6) negotiations which preceded the Dillon round proper.

in any case favoured the domestic crushing and refining sectors. To the United States it seemed like a useful concession, but the US was generally unhappy with the outcome on other products. It left over from the Dillon round 'unsatisfied negotiating rights' to be used in later trade talks.

What followed is well known. High wheat prices set for the new CAP (a political compromise which was made to satisfy German rural interests) drove feed demand to maize - much of it imported from the United States. When grain prices were harmonised in the European Community under the so-called silo system (moving maize prices up to meet the higher wheat prices), feed compounders switched to non-grain feeds. The pressure was particularly strong in Germany and the Netherlands, where 'green' exchange rates ensured higher cereal prices than in other parts of the EC. Compounders found that a mix of soybean meal and cassava chips (from Thailand) made an excellent substitute for maize, barley and wheat.

Soybean sales to the European Community boomed over the 1970s and continued high when EC corn market imports slowed down. The hole in the dike was letting in a flood of imports. Later, when the EC entered into a bilateral agreement with Thailand on a limit to cassava imports, other starchy feed ingredients, such as citrus pulp, maize germ meal and distillers' dried grains, began to be used to replace cereals in feed rations - many imported from the US. Corn gluten feed, a by-product of the corn wet-milling process in the US, became the ingredient of choice in the 1980s. [5] The European Community classifies all these starchy products as 'cereal substitutes': the high-protein meals are, of course, better characterised as 'cereal complements'. [6]

The predominant US view is that the Dillon round concession was negotiated in good faith, and the fact that it has turned out to be valuable is not through any fault of the US. It is one of the few things that the United States has 'won' in the GATT: to let the concession slip would undermine the domestic credibility of that organisation. Soybean and non-grain feeds are among the few market-orientated sectors in world trade. An assault on this trade by the European Community would represent a major set-back for the aims of liberalisation. And, until recently, the oilseed sector in the EC was a model of market orientation, to be copied rather than destroyed in a fit of rebalancing.

The conflict simmers on three fronts. The Community proposal to institute a fats and oils tax, often rejected by the Council of Ministers but never quite abandoned by the Commission, is seen in the United States as a scarcely disguised tax on soybean imports. The fact that the tax would apply to

---

5   The wet-milling of corn is used to produce high-fructose corn syrup (isoglucose), a substitute for cane and beet sugar, and ethanol, used as an additive to gasoline.

6   The distinction is not always clear. At times feed ingredients high in protein have been used for their carbohydrate content.

domestic as well as imported oils cuts little ice. The fats and oils tax to be effective would undoubtedly have to reduce consumption of soybean oil, benefiting the market for butter and perhaps olive oil. US opposition to the tax has been made abundantly clear, and this has perhaps influenced the member states (the United Kingdom and West Germany) that have consistently opposed it within the EC.

The attempt in 1985 to limit imports of corn gluten feed from the US - following the 'success' of the voluntary export restraint agreement with Thailand - also received a rebuff from the US. Why enter voluntarily into an agreement to limit a booming export market when the open market was due to a firm commitment in GATT? [7] At least, if the negotiations were undertaken in the GATT, the US could have expected to be offered compensation. Unilateral action of this kind was ruled out of the question in Washington.

The third 'front' of this conflict has been the growth of oilseed production in the EC, promoted in part by subsidies to crushers who use domestic oilseeds (and pay at least a minimum price). Production of soybeans has increased in Italy, rapeseed has become a popular crop in Northern Europe, and sunflower production has expanded in the South. Once again, the US view is clear-cut: the EC is trying to promote the production of one of the few products that it has actually allowed in from abroad. Nothing could show more clearly that the true aim of the CAP is to eliminate imports and close the door to the rest of the world.

The threat to US soybean export did not go unnoticed. The United States Soybean Association lodged a complaint against the EC oilseed programme, and the US Administration successfully pushed for a GATT investigation of the case.

The US argued that the GATT binding on soybeans and products was being offset by the domestic subsidies. They also complained about the way in which such subsidies were paid (to crushers) and the fact that they varied with world prices. The GATT panel ruled against the EC, raising the interesting speculation that existing GATT rules may actually be useful to control domestic policies if applied with rigour. The EC will have to adjust its policy in the light of the panel ruling, just as the US will have to modify its sugar policy following another adverse panel decision. [8]

---

7   The duty on corn gluten feed had also been bound at zero in the Dillon round.

8   The complaint was brought by Australia, which argued that the US was changing its agreed schedule of trade restrictions in a way that impaired their rights under the GATT. An attempt by the EC to challenge the 1955 US waiver as it applied to sugar did not succeed.

Beef Imports

A recent trade conflict that illustrates well the predominant United States view of European Community agricultural policies is that over beef imports. The EC is not a major market for US beef. Most exported beef goes to Japan, a market which seems to appreciate the grain-fed beef with the marbled texture produced on US feedlots. The US in turn imports considerable quantities of grass-fed manufacturing beef from Australia and Central America, together with some cooked meat from South America. The EC has moved from being a steady importer of South American and Australian beef to one of the world's largest beef exporters. Beef imports are controlled by both a tariff (of 20 per cent) and by variable levies, to protect the domestic market price. However, beef is still imported under a number of schemes for preferential access, such as levy-free and tariff-free quotas. The US has a levy-free quota for high-quality ('Hilton') beef, accounting for some $10 million annual sales. But much more important has been the growth in the market for 'variety meats', comprising beef offals used in making pies and sausages in Europe. This market has reached $100 million, and provides a valuable outlet for beef by-products not commonly used at home. It is this trade that was threatened by the import ban.

In December 1985, the European Community decided to ban the use of anabolic hormones in livestock production. The ban followed rising consumer concern over the health effects of such hormones, spurred by the well-publicised incidents in Italy involving the use of a synthetic oestrogen, diethylstilboestrol (DES). DES had been banned in the EC (and the US), but consumer confidence in the safety of cattle feeding and in the ability of the regulatory authorities to control such practices was shaken. The hormone ban, applicable at first to domestic production and then extended to exporting countries in January 1989, was ostensibly a reaction to this consumer pressure. The United States reacted to the loss of the trade in beef and offals by imposing a 100 per cent tariff on an equivalent value of EC exports to the US. A European Community counter-retaliation was announced, but has so far not come into force, pending bilateral negotiations.

The dominant US view of the beef hormone ban sees the Commission's actions as essentially of a protectionist nature, bowing to pressure from cattle producers and being concerned with the build-up of intervention stocks of beef. No one doubts that there have been consumer concerns over the health effects of hormone use in cattle. But there is no credible scientific evidence that hormone treatment of beef cattle, under proper conditions, leaves any harmful residue in the meat sold to consumers. The United States Food and Drug Administration tests artificial growth hormones before approving their use in livestock production, and the USDA has a test programme to check for chemical residues in meat. Therefore, the role of public authorities in the EC should have been to educate consumers on the matter, and to make sure that producers followed accepted practices. Consumer views don't seem to be heeded when it comes to heavy taxes on food: why should they suddenly be so important when the issue is hormones in beef production? The answer seems clear. Allow the issue to be settled by scientific evidence and not by uninformed public sentiment.

UNITED STATES VIEW ON CAP REFORM

United States and European Community officials can agree on at least one thing: that the CAP threatened the stability of the Community as a whole in the early 1980s as a result of soaring budget costs. The EC had to deal with the issue of putting agricultural finance on a sure footing and preventing it from precluding other desirable activities, such as fighting unemployment, protecting the environment, developing technology, enlarging the Community and completing the internal market. The CAP had to be reformed (that is, brought under control) before it did serious damage to the Community. The US saw floods of subsidised exports, shrinking EC markets for imports and a succession of trade squabbles. Reform in the CAP would surely imply drastic changes in support prices or sharp policy shifts to remove production incentives. How far reform has proceeded can be argued among reasonable observers, but, from the predominant US viewpoint, the process has been woefully inadequate.

Policy reform in the context of the CAP commonly refers to the introduction of milk quotas in March 1984, and their strengthening in December 1986; and to the adoption of the 'stabilisers' package in February 1988. [9] The changes in milk policy were not of great importance to United States trade interests. The US is roughly self-sufficient in dairy products. It imports speciality cheeses from the EC, while disposing of American cheese under domestic and foreign concessional programmes. The US disposes of limited amounts of skim-milk powder (non-fat dried milk) abroad, mainly through overseas aid programmes. There is little trade in butter, although there are occasional domestic surpluses. CAP reform in the dairy sector has clearly reduced EC dairy capacity and firmed up world dairy product prices, but the effect on the US has been minimal. [10] It is seen as a domestic reform, aimed at cutting budget costs. The introduction of quotas, as an alternative to the politically more difficult price cuts, is not regarded as a great contribution to far-sighted policy-making. Though the US and many other countries use variations of quotas in controlling the dairy industry, it is seen as moving in a direction counter to that of deregulation. Since quotas once introduced are hard to remove, the net result may be to reduce the chances for true liberalisation in the future.

---

9   For a discussion of these reforms see Moyer and Josling (1990) and Chapter 16 of this book.

10   The United States had its own dairy scheme in recent years, the Dairy Herd Replacement Program, which significantly (if temporarily) reduced the dairy cow numbers.

The relative lack of impact of EC dairy quotas on US trade may explain why that aspect of reform is not widely recognised across the Atlantic. The same cannot be said about the 'stabilisers' programme, and the introduction of set-asides and direct income payments that together make up the latest CAP reform package. How the EC tackles its grain surplus and the growing expenditure on oilseeds payments is of direct interest to the United States. The prevailing US impression of this package is that it is a small start to correcting a large problem, and that the applause should be withheld at least until there is evidence of further similar measures. To tout it, as the Commission does, as a major change in EC agricultural policy, a significant contribution to the adjustment in world agriculture, and a step which shows to others the way of the future is to stretch credibility to the limit.

That stabilisers programme itself is a fairly modest quasi-automatic device for linking price levels to output at the Community level. 'Guarantee thresholds' have been around for much of the 1980s, but were generally regarded as ineffective. The 1988 programme extended their use to most major products, including cereals, and made more certain the penalties for over-production. Large-scale cereal producers now face an additional co-responsibility levy of 3 per cent (in addition to the 3 per cent levy already in place), which will only be reimbursed if output falls short of the maximum guaranteed quantity (MGQ). The levy is translated into a fall in the intervention price the following year. Assuming that the Commission does not negate these automatic price reductions at the annual price review, the price restraint could have an effect on output. But, set against productivity changes in the cereals sector, the fact that small farmers are exempt from the co-responsibility levy (as is grain used for feed on farm) and the generally upward movement of prices due to the green currency system, the restraint is modest. And, since the cereals stabiliser only lasts for three years, there is no guarantee that any permanent change has been achieved.

Oilseed and protein crops are included in the stabilisers programme, in part because of the steady rise in support costs for these products, and in part to anticipate any possible shift out of cereals towards these commodities. [11] The net effect on the price of rapeseed and protein crops (field peas and beans) is expected to be minimal, although the programme may help to inhibit the growth of soybean production in Spain (Australian Bureau of Agricultural Resource Economics 1989). It could be argued that the United States has not shown enough appreciation for this attempt to control the growth of oilseed production, but from the 'outside' the policies look less than dramatic.

The introduction of set-asides in the European Community should have triggered a sympathetic response. The United States has been using set-asides as a major aspect of policy since the 1930s, and in recent years has made

---

11 One might have expected the Community to have welcomed any shift out of cereals, since that presumably was the rationale for price restraints in that sector.

participation in the major price support programmes conditional upon compliance with mandatory acreage adjustment. The reaction to EC set-asides has hardly been enthusiastic. It is apparent that the national schemes presently on the books will not be sufficiently attractive to entice many acres out of production. The Commission's estimate of one million hectares (and 3.5 million tonnes of grain) is considered optimistic. Individual countries have varying degrees of commitment to the scheme, which was pushed by the German Government as an alternative to the stabiliser price cuts. Since national governments will have to pay up to 65 per cent of the cost, their incentives to encourage participation are relatively weak. Add to this the fact that even the most extensive set-aside programmes in the US seem to have a relatively modest impact on production, as farmers find ways of idling acres without losing output, and the attitude is cautious at best and scornful at worst.

One aspect of the EC's reform package that might seem to be high on the popularity charts is the element in the socio-structural measures which allows for direct income assistance to farmers. Farmers earning significantly below the national average farm income level could receive up to 1500 ECU per year, with the member states covering up to 70 per cent of the cost. At present this looks to be a relatively minor add-on to the range of structural programmes that are available. Much depends on how the member states decide to administer the scheme. Funding for regional and structural programmes as a whole, both agricultural and more general schemes, has been increased sharply. But there is little indication that the Commission is considering the move of any major part of the present price support transfers to direct income payments of a decoupled nature. From the Community's perspective, the introduction of a scheme of this type is significant. It establishes the principle and allows the political process to get used to the idea: it may also provide useful experience in administration. To the US observer, the scheme cannot be very significant if it is so small and under-funded.

The muted US response to what is often regarded in Europe as a major change in policy may seem churlish. In part this response is conditioned by the experience of years of well-meaning efforts by the Commission to make significant changes in the CAP, only to be rejected by the Council of Ministers. In part it reflects the different viewpoint, that results count and the amount of political contortions needed to achieve the decision is not of interest. In part, it is a function of the present stage of multilateral negotiations, where to recognise that the European Community has made major changes may itself squander negotiating capital.

UNITED STATES-EUROPEAN COMMUNITY PERCEPTIONS IN THE GATT ROUND

The start of the Uruguay round of GATT negotiations, in September 1986, provided a new setting for US-EC discussions on agricultural policy. United

States reaction to the Community position in the GATT talks illustrates many of the points made earlier, but, in the context of a formal negotiation, these perceptions begin to take on an added significance. This is particularly true of the current round, where by mutual agreement domestic agricultural policies are 'on the table'.

The first step in the Uruguay round negotiations on agriculture was to elicit from the major participants their ideas for the conduct of the talks. This prompted the US Administration to table a somewhat radical proposal to eliminate, over a ten-year period, all trade-distorting price support for agricultural commodities. The Cairns Group (of smaller agricultural exporters, led by Australia and Canada) also proposed liberalisation, but seemed to offer a less abrupt transition to this state of grace. The Community countered with a paper which emphasised short-term action to correct the depressed situation on world markets, which would then be followed by 'significant' reductions in levels of support and some rebalancing of that support.

The year 1988 saw intensive negotiations on these proposals in Geneva. The US wanted the Community to sign on to the notion of long-term trade liberalisation (the 'zero' option): once that had been agreed then shorter-term issues could be addressed. The Community saw no reason to move so far beyond the Punta del Este agreement (for 'greater liberalization of trade in agriculture') and doubted that the US was really serious in its proposal. The stand-off lasted throughout the year, and caused the 'failure' of the Mid-Term Review in December 1988, the ministerial meeting that was to have provided the agenda for the remainder of the round. In the event, an agreement was put together in April 1989 in Geneva which has allowed negotiations to proceed.

To the US, the EC reaction to its proposal confirmed the view of the Community as a protectionist group. Why would the EC not come back with a counter-proposal for long-run policy reform, if the zero option was too drastic? Would it not assist the economic development of Europe, as well as its political cohesion, to scale down or remove the element of protectionism in the CAP? Farm incomes could still be supported by a variety of production-neutral programmes, and research and extension programmes, together with development assistance and food aid, would not be touched. What would go would be the troublesome set of market policies which had proved so costly. Surely the European Community could not resist the logic of the argument for agricultural market liberalisation!

The EC response was deemed unsatisfactory in three other regards, besides not taking seriously the US proposal. First, the Community proposal had unmistakable traces of a market-sharing philosophy, anathema to US competitive instincts. Even if market shares were only suggested as a way out of temporary problems in a few markets, the signs were ominous. It is widely remembered in the US that the EC went into the Kennedy round with notions of world commodity agreements (as spelled out in the Baumgartner/Pisani

Plan). The view of the European Community as favouring market management over liberalised trade is pervasive. For the European Community to propose such solutions in the GATT was regarded as further evidence.

A second concern was the hint that the GATT round would be used as a way of correcting the relative price distortions that had dogged the cereals, oils and fats markets. If reducing protection overall could allow for some increase in protection, then the genie would be out of the bottle. The soybean and corn gluten feed markets would disappear under some broad cloak of GATT respectability, and the US farmer would once again have been duped by the clever Europeans. In part for this reason, the US has emphasised in its own early submissions to the GATT the notion of 'country plans' which would have to be acceptable to trading partners. Support reduction would not be tied just to some formula and based on abstruse calculations: it would be tangible and transparent, and allow for the calculation of mutual advantage.

The third caution raised by the Community position in the GATT related to the notion of credit for actions taken. The trade ministers at Punta del Este decided that it would be wise to allow countries which constructively modified domestic policies in advance of any GATT agreement to claim credit for their actions. The EC has been asking for credit for its reform programme, on occasions even suggesting that the 1984 dairy policy changes be counted. Such requests for credit have two implications: they take the pressure off the European Community to make further changes in the CAP in advance of an agreement (and possibly not even then, if world prices remain firm); and they allow the EC to shift the focus on to others, and to request that they too make a serious start on policy reform.

Neither of these implications is particularly acceptable to the United States. It seems implausible that a GATT agreement is saleable to US domestic interests if it does not include some major liberalising moves by the Community. The politics of 'equal degrees of pain' will ensure that the EC cannot just live on credit for the next few years, even if an indicator could be found that would validate a claim for such credit. Nor is it likely that the US can agree to undertake domestic reform as a way of catching up with the EC.

The US reaction to the notion that it should undertake unilateral reforms seems rather to have been to 'keep the pressure on' the EC through an expanded Export Enhancement Programme and a relaxed set-aside. This is not building up credit, it is accumulating debts. The US might still be able to adopt a radical liberalisation of trade, which it would claim was inspired by the tactics of pressure, in conjunction with other countries. It seems most unlikely to be persuaded to take the first steps alone.

Developments since April 1989 have not resolved any of these issues. The US modified its own position, but not in a way that appealed to the EC negotiators. Instead of phased reductions in support, allowing a degree of policy choice by individual governments, the US began to be more explicit as to which policy instruments were to be modified. By the time of tabling of the

US comprehensive proposal at the end of 1990, the US was asking for conversion of the EC variable levy to a fixed tariff, the phasing out of export subsidies and the abolition of all price-related domestic measures such as intervention buying and direct payments. A more comprehensive attack on the CAP could not have been imagined. The EC, meanwhile, had embraced the notion of phased reductions in overall support (though not to the same extent as the US had suggested in 1987) and shunned the notion of singling out individual instruments of policy. The EC comprehensive proposal did offer to discuss limited movements towards the tariffication of levies, making them less variable, but only in return for agreement to 'rebalance' support among commodities.

These developments illustrate the points made earlier. The original US proposal, in July 1987, appears a product of the 'ERS view', that all countries have become locked into trade-distorting domestic policies and need the guidance of international agreements to move to more acceptable domestic methods of income support. The extensive modelling in the OECD, the USDA and other agencies provided evidence of the soundness of this approach. Moving together, countries could avoid some of the costs of unilateral policy reform: remaining income support programmes would in turn be both more efficient in domestic terms and less disruptive to others.

It appears that this benign view of the policy process ran aground on the rocks of producer self-interest at some time during 1988. The models appeared to show that US farmers (and EC farmers) would stand to lose considerably from trade liberalisation, to a greater extent than they could hope to recoup from compensatory payments and decoupled subsidies. Moreover, commodity-based lobby groups in the US began to see their influence being weakened by 'across the board' formula approaches to policy reform. The 'FAS view' began to re-emerge in the US position. Tariffication (together with a ban on export subsidies) would finally declaw the CAP: a fixed tariff, negotiated down over time, would remove any role for EC domestic policy in the cereals market. This would be a tangible outcome of the Uruguay round, saleable at home, and superior to any non-policy-specific agreement to reduce overall support levels.

CONCLUSION

The Uruguay round negotiations are proving a focus for the resolution of many of the US-EC agricultural trade conflicts. But even a constructive agreement to impose some agreed disciplines on domestic agricultural policies will not change overnight the US perception of the CAP. The agreement itself will no doubt contain complex transitional rules and allow plenty of scope for contentious interpretation at a later date. The EC will probably wish to turn attention to internal matters at the conclusion of the Uruguay round, making it likely that overseas interests will tend to be ignored. Other trade issues will arise even if the traditional battleground of cereals and oilseeds is calm. The beef-hormone conflict may prove the tip of a rather large iceberg of trade concerns arising from consumer reactions to biotechnology. Whether GATT rules can keep ahead of these trade tensions seems doubtful: perhaps issues have to be played out as conflicts before international agreements can be framed.

A failed Uruguay round raises a more disturbing spectre. If cracks appear in the multilateral trading process, blame will undoubtedly be placed on agriculture and more specifically on the intransigence of EC (and Japanese) politicians. Under these circumstances one could expect a period of more aggressive policy actions. The 'FAS view' would appear vindicated. The EC will have shown that it is unable to put broader economic interests ahead of narrow sectoral concerns. The 1992 programme for completion of the internal market could also take a sharp turn in the direction of protectionism if the Uruguay round were to fail. The result would be to reinforce the view of the EC as a bastion of protectionism. Moderate voices would be stilled and the multilateral process discredited. A new chapter in the annals of EC-US agricultural relations could be just beginning.

REFERENCES

Australian Bureau of Agricultural Resource Economics (1989) *The 1988 EC Budget and Production Stabilisers,* Discussion Paper No. 89.3, Canberra.

Josling, T. *et al.,*(1990) *Report of the Task Force on the Comprehensive Proposals for Negotiations in Agriculture,* International Agricultural Trade Research Consortium, Washington, USA.

Moyer, H. Wayne and Josling, T. (1990) *Agricultural Policy Reform,* Harvester Wheathseaf, Hemel Hempstead.

Newman, M., Fulton, T. and Glaser, L. (1987) *A Comparison of Agriculture in the United States and the European Community,* USDA/ERS, FAE Report No. 233, Washington DC, October.

USDA (United States Department of Agriculture) (1988) *Western Europe: Agricultural and Trade Report,* ERS, Situation and Outlook Series, Washington, June.

CHAPTER 15

THE CAP, CUSTOMS UNIONS AND THE MEDITERRANEAN COUNTRIES[1]

Alan Swinbank and Christopher Ritson

INTRODUCTION

Chapter 9 considered the impact on agricultural trade between EC member states of the Customs Union upon which the Common Agricultural Policy is based. It also commented upon the implications of this Customs Union for agricultural trade with non-member states. Among the most affected have been the countries bordering the Mediterranean Sea. Conscious of this, the Community has developed a series of preferential trading relations with most of the Mediterranean countries. In some cases, these agreements have specified the eventual creation of a customs union between the Community and the country concerned, which raises the question of how such an agreement would differ, in so far as agricultural trade is concerned, from actual membership of the Community.

This chapter therefore reflects upon some of the implications of the EC's preferential arrangements with Mediterranean states (referred to hereunder as the EC's Mediterranean Associates), particularly as they affect trade in agricultural products, in the aftermath of the Iberian Enlargement.

THE MEDITERRANEAN AGREEMENTS

Over the years, the EC has negotiated preferential trade agreements of various sorts, and of varying degrees of liberalism, with all the Mediterranean states except Libya and Albania. The accession of Portugal and Spain to the EC on January 1, 1986 meant that these agreements had to be revised, but it was not until the November 25-26,1985 meeting of the Foreign Affairs Council that a

---

1 This chapter is adapted from Swinbank and Ritson (1988). The authors have jointly acted as consultants to the Government of Cyprus on the agricultural aspects of the Cyprus: EEC Association Agreement. Many of the ideas expressed in this chapter emerged as a result of discussions with officials, and others, in Nicosia, Brussels and London.

mandate was agreed for 'the Commission to open negotiations with the Mediterranean countries with a view to adapting the Co-operation and Association Agreements following enlargement' (Council 1985 p 4).

In April 1986, however, Spain, dissatisfied with the arrangements made for exports of fruit and vegetables from the Canary Islands to the EC(10), in the Council blocked a Commission request to improve the proposed concessions to the Mediterranean Associates (*Agra-Europe* April 25, 1986 p E/3). In October 1986, therefore, in order to meet these Spanish objections, the Council approved a revised mandate 'to enable the Commission to hold the final stage of negotiations with Mediterranean third countries', as well as approving 'guidelines for adaptation of the arrangements applicable to the Canary Islands' (Council, 1986, p 1).

A number of the agreements with the Mediterranean Associates were settled in the early summer of 1987. Thus the revised agreement with Tunisia was signed on May 26, 1987; those with Algeria and Egypt on June 25; and those with Jordan and Lebanon on July 9, 1987 (see *Official Journal of the European Communities OJ/L297*, October 21, 1987 for the legal texts). [2] The revised agreement with Cyprus, together with an agreement on implementation of the second stage of the Cyprus: EEC Association Agreement, was signed in October 1987, and will be discussed more fully below.

The Israeli agreement was more problematic. In the spring of 1987 it was reported that Israel's Foreign Minister had had talks in Madrid

'in an attempt to resolve a bilateral dispute holding up the implementation of Israel's draft agricultural access agreement with the European Community...Spain refused to initial the draft treaty last December until Israel, in turn, signs a "protocol of adaptation" reducing its tariff barriers on Spanish industrial exports' (*Financial Times* April 7, 1987).

Later, a dispute over the Israeli treatment of produce originating in the occupied territories was reported to be the main impediment to the conclusion of the revised Israel: EC agreement (*Financial Times* November 17, 1987); which was, however, finally signed on December 15, 1987 and came into force on December 1, 1988 (*OJ* L327 November 30, 1988).

---

2   The historical progress of the Tunisian agreement (Additional Protocol to the Co-operation Agreement between the European Economic Community and the Republic of Tunisia) is indicative of the timescale involved: it was 'initialled' by the Commission and Tunisia in December 1986 (*Bulletin of the European Communities,* 1986, No. 12, point 2.2.18), signed by the Plenipotentiaries of the two parties, the Council of the European Communities and the Government of the Republic of Tunisia, on May 26, 1987, approved by the European Parliament on September 16, 1987, and finally adopted by the Council on September 28, 1987 (*OJ* L297). Subsequently, it entered into force on November 1, 1987 (*OJ* L309).

This is not the place to discuss in detail the form, and legality, of the respective Co-operation and Association agreements between the Mediterranean states and the EC. Suffice it to say, their status in GATT is not wholly clear. Article XXIV of GATT makes provision for the establishment of customs unions and free trade areas: in particular paragraph 5 provides:

'5. Accordingly, the provisions of this Agreement shall not prevent, as between the territories of contracting parties, the formation of a customs union or of a free-trade area or the adoption of an interim agreement necessary for the formation of a customs union or of a free-trade area; Provided that:...(c) any interim agreement...shall include a plan and schedule for the formation of such a customs union or of such a free-trade area within a reasonable length of time' (The General Agreement on Tariffs and Trade; as reprinted in Dam (1970) p 432; or McGovern (1986) pp 571-572).

Some years ago the GATT arbitration panel, established to examine EC preferences for Mediterranean citrus products, noted that

'...the agreements currently in force between the Community and the Mediterranean countries concerned, under which EC preferences on citrus are granted at this time, have been presented to the GATT by the parties as interim agreements leading to the formation of a customs union under Article XXIV (Cyprus, Malta and Turkey), as interim agreements leading to the formation of a free-trade area under Article XXIV (Israel and Spain), or as agreements comprising a free-trade area obligation on the part of the EC under Article XXIV but [with] no reciprocal commitments by the other parties consonant with Part IV (Algeria, Egypt, Jordon, Lebanon, Morocco and Tunisia)' (GATT 1985, p 13).

Part IV of GATT deals with 'Trade and Development', and establishes the general principle that 'The developed contracting parties do not expect reciprocity for commitments made by them in trade negotiations to reduce or remove tariffs and other barriers to the trade of less-developed contracting parties' (Article XXXVI (8)).

Hine (1985) pp 44-45 comments:

'The use of Article XXIV to legitimise the formation of the EEC set a precedent which the EEC has exploited repeatedly in the elaboration of its trade relations with third countries. By this means the EEC has been able to spin a worldwide web of preferential agreements whilst claiming to adhere to the rules of GATT.'

Gaines *et al.* (1981, p 433) have taken a similarly robust line, in flatly declaring that 'The Mediterranean agreements...are in fact preferential agreements and not interim agreements, and thus do not meet the conditions of Article XXIV.'

McGovern (1986, p 267), however, has suggested that the EC may be justified in invoking Part IV of GATT in legitimising the agreements with the Maghreb and Mashraq states. Spain is now an EC member state, and Turkey, Malta and Cyprus have applied for membership; but this does leave open the interesting question of what form a free trade area with Israel, and customs unions with Cyprus and Malta, could take - particularly in respect of the provisions of the Common Agricultural Policy (CAP).

For Israel and Malta the interim agreements, as Gaines *et al.* (1981) indicate, may prove to be of indefinite duration; but for Cyprus customs union [3] is now the EC's publicly stated objective. This has not always been the case: as Tsardanidis (1984) remarks:

'The customs union was vaguely mentioned in the Association Agreement drafted in 1972. Only in the joint declaration by the contracting parties concerning Article 2 of the agreement [is it] mentioned that "the Republic of Cyprus envisages the progressive establishment, during the course of the Agreement, of a customs union with the EC" as well as in the preamble of the agreement... But also the concept of a customs union between the Community and a third non-member is not as yet defined in theory or in practice.'

CUSTOMS UNION

Article XXIV(8) of GATT is clear:

'8. For the purposes of this Agreement:

(a) A customs union shall be understood to mean the substitution of a single customs territory for two or more customs territories, so that, (i) duties and other restrictive regulations of commerce...are eliminated with respect to substantially all the trade between the constituent territories of the union or at least with respect to substantially all the trade in products originating in such territories, and (ii)...substantially the same duties and other regulations of commerce are applied by each of the members of the union to the trade of territories not included in the union...'

Twenty years ago, Dam (1970, pp 275-276) remarked:

'Today it is clear that if a single adjective were to be chosen to describe Article XXIV, that adjective would have to be "deceptive". First, the standards established are deceptively concrete and precise; any attempt to apply the standards to a specific situation reveals ambiguities...Second, although the rules appear to be based on economic considerations, the underlying principles make little economic sense. Third, the dismaying experience of the GATT has

3   Although, as noted, membership applications have now been made by both Malta and Cyprus.

been that, with one possible exception [the United Kingdom/Ireland Free-Trade Area], no customs-union or free-trade-area agreement thus far presented for review has complied with Article XXIV, yet no such agreement has been disproved.'

Agriculture, but more particularly the CAP, presents particular problems. Put bluntly, the question is: 'Is the EC's CAP separate and distinct from its customs union, or is the CAP merely the mechanism whereby the customs union in agricultural products is effected?' Article 38 of the EEC Treaty is itself rather ambiguous and does admit of a distinction: paragraph 1 establishes that the common market extends to agriculture and trade in agricultural products, but paragraph 4 goes on to specify that 'The operation and development of the common market for agricultural products must be accompanied by the establishment of a common agricultural policy among the Member States.'

The view that the CAP and the customs union are two separate EC policies would be in accord with the previous EC negotiating stance in GATT: that the CAP was a domestic agricultural support policy and not a trade mechanism, and hence not negotiable. Thus, at the time of the Tokyo round, the official EC view was that the CAP's 'principles and mechanisms should not be called into question and therefore do not constitute a matter for negotiation' (as quoted in Harris 1977, p 39).

Such a stance, however, would appear to have been eroded by the Punta del Este Ministerial Declaration that, in the Uruguay round,

'Negotiations shall aim to achieve greater liberalisation of trade in agriculture and bring all measures affecting import access and export competition under strengthened and more operationally effective GATT rules and disciplines...' (*Financial Times*, September 22, 1986).

If it is held that the CAP is separate from the customs union (or free trade area) then, in a customs union or free trade area between the EC and a third country, customs duties and quantitative trade restrictions on agricultural products would be eliminated, but variable import levies (on cereals, sugar, milk products and so on) and reference (minimum import) prices on wine and fruit and vegetables would remain. It might be noted that the EC's free trade agreements with the EFTA countries exclude agriculture altogether.

If the CAP is merely the mechanism whereby the customs union is effected in the agricultural sector, then customs union between the EC and a third country implies the adoption of the CAP (or at least CAP-compatible mechanisms) by that third country; and for this, and other reasons, it is far from clear how such a state differs from Community membership. It may be that the Mediterranean states would welcome an agreement which removed customs duties and reference prices from their exports of fruit and vegetables to the EC, but the quid pro quo could involve the importation of cereals, sugar, milk products, beef, and so on, at CAP prices from EC member states, or over the common variable import levy from non-customs union sources.

The latter comment raises a further question, which is rarely addressed in the theoretical literature on customs unions, although the same issue does not arise in a free trade area: how are the receipts from the common customs duties (or common variable import levies on CAP products) to be treated? The approach adopted by the EC, foreshadowed in Article 201 of the Treaty of Rome, has been to treat import duties, and variable import levies, as the 'own resources' of the Community. This was not just to raise revenue for the Community, but was also a recognition of the 'Rotterdam effect' in that import duties collected in one member state might in fact be paid by consumers in another member state. Similarly, there has until recently been a view that the CAP should be commonly funded, in that export refunds or intervention mechanisms paid for in one member state are part of a 'seamless cloth' supporting CAP prices throughout the Community.

The requirement, in Article XXIV, that the customs union (or free-trade area) should involve the elimination of customs duties and other restrictive regulations of commerce on substantially all the trade between the constituent territories is also ambiguous. What is meant by substantial, and how is it to be assessed? A figure of 80 per cent is sometimes cited (following Dam (1970), pp 279-280, who uses 80 per cent as a hypothetical example?); but if there is an 80 per cent rule how is it to be interpreted? Is it to be a common list which comprises both 80 per cent by value of the EC's exports to its customs union partner, and 80 per cent of the partner's exports to the EC; or does it suffice if the list covers 80 per cent of the total trade by value between the EC and its customs union partner, involving, say, 75 per cent of the trade in one direction and 85 per cent of the trade in the other? What happens if trade patterns change, which is to be expected in any case as a result of the partial elimination of trade barriers? How is the 80 per cent rule to be interpreted in the light of the EC's tangled web of free trade agreements (with EFTA for example) and customs unions? Is it possible for the EC to have a customs union with more than one Mediterranean state, involving different traded goods under the 80 per cent rule?

The 80 per cent rule, if it exists, does, however, seem to offer the prospect of a differential approach to agricultural products in any future full customs union or free trade area between the EC and a Mediterranean state. If fruit and vegetables are among the products covered by the agreement, then duty-free access to the EC for the Mediterranean state's export goods could be secured (remembering that this also involves duty-free access for similar EC goods into its partner's market, perhaps at a different time in the season); but products subject to the full rigour of the CAP, and imported into the Mediterranean state, such as cereals and dairy products, could be excluded.

Although the EC and its Mediterranean Associates have as yet made little progress in their quest for fully operational customs unions or free trade areas, this is not to say that the EC will always be a reluctant partner. It is possible (if not probable) that the EC's enthusiasm for fully fledged agreements might, at some future date, be fired. A number of reasons could be adduced:

a) the political importance of particular states in the Mediterranean Basin might prompt the EC into forging closer economic links;

b) to 'save face' in GATT, it might be necessary to engage in a genuine movement towards customs union or free trade area status with one or more of its Mediterranean Associates;

c) if a wider Mediterranean initiative were to be taken, it might deflect attention from Turkey's embarrassing application for membership;

d) a breakdown in world trading relations in agricultural goods, notwithstanding the good intentions expressed at the outset of the Uruguay GATT round of trade negotiations, might encourage the EC to secure outlets in the Mediterranean Basin for its farm surpluses. Table 15.1 indicates that 11-12 per cent of EC(12) agricultural exports are destined for Mediterranean states, including significant quantities of cereals and livestock products. In comparison, only 5-6 per cent of EC(12) agricultural imports come from the region. The Southern Mediterranean states are in fact heavily dependent upon cereal imports, though not just from the EC, for animal feed and human food; and the indications are that they will continue to be so dependent into the foreseeable future (Genazzini and Hörhager 1987);

Table 15.1:   EC(12) Agricultural Trade* with the Mediterranean Basin**

	Imports (m ECU)			Exports (m ECU)		
	1980	1986	1987	1980	1986	1987
Extra-EC(12) with:						
World	44,130	52,802	50,832	20,215	28,804	28,435
of which:						
Mediterranean Basin	2,120	2,816	2,804	3,192	3,449	3,209
of which:						
meat & live animals	36	26	31	294	436	444
milk & eggs	3	6	8	527	546	551
cereals	12	10	14	1,010	827	578
animal feed	11	14	16	148	225	158
sugar & honey	12	20	24	356	291	393
oils & fats	122	20	42	282	350	315
fruit & vegetables	1,228	1,696	1,661	166	186	178

* SITC 0, 1, 21, 22, 232, 24, 261-265, 268, 29, 4, 592.11-12
** Malta, Turkey, Morocco, Algeria, Tunisia, Libya, Egypt, Cyprus, Lebanon, Syria, Israel and Jordan.
Source: Commission (1989), pp 153-154, 156.

e) the dramatic changes in Eastern Europe are leading the Community to consider appropriate arrangements for its trading relations with the countries concerned, and the concept of a customs union could provide a model for an appropriate relationship which fell short of full membership.

Despite fears that the democratic revolutions in Central and Eastern Europe, and the unification of Germany, would divert the Community's attention from the Mediterranean, during the course of 1990 there were some signs of a renewed initiative towards the South. Thus, the Commission submitted to the Council a communication entitled *Redirecting the Community's Mediterranean Policy* (as reported in Commission (1990) and Economic and Social Committee (1990)), which was discussed by the Council on several occasions; and in October 1990, on the initiative of Italy, which held the presidency of the Council of Ministers in the second half of the year, five North African foreign ministers and an observer from Malta, met their Italian, French, Spanish and Portuguese counterparts in Rome to discuss launching a new exercise in regional collaboration (*Financial Times,* October 10 and 11, (1990)). According to the Economic and Social Committee (1990):

'The proposal to relaunch Mediterranean policy forms part of a new "policy of neighbourly relations" aimed at strengthening the Community's links with its closest neighbours: EFTA, Eastern Europe and the Mediterranean.'

THE CYPRUS : EEC ASSOCIATION AGREEMENT AND THE CAP

In October 1987 the EEC and Cyprus signed a 'Protocol laying down the conditions and procedures for the implementation of the second stage of the agreement establishing an Association between the Republic of Cyprus and the EEC and adapting certain provisions of the Agreement', which came into force on January 1, 1988 after ratification by both parties (*OJ* L393, December 31, 1987). In part, the Protocol amended the Association Agreement as a result of the Iberian Enlargement, and in part it was concerned with the 'implementation of the second stage' of the Association Agreement (Association Council 1987; and *Cyprus Bulletin,* 25(22), October 29, 1987). Thus the Protocol provides for 'transition to the second stage of the Agreement, which should gradually lead to full completion of a customs union between the EEC and Cyprus after a period not exceeding 15 years' (Association Council 1987, p 2).

This second stage is, however, split into two phases. The first of these phases lasts ten years, and is defined in the Protocol. The second phase of the second stage only comes into force by decision of the Association Council, after defining the 'various measures required for the completion of the customs union', and is expected to last no more than five years, following which the customs union will be attained. This second phase has still to be defined.

Over the ten-year period of the first phase of the second stage, Cyprus will gain improved access to the EC market for those fruits and vegetables and wines already granted concessions, and some others; Cyprus will make limited concessions to the EC on agricultural products, notably sugar; and Cyprus will progressively adopt the Common Customs Tariff for products included within the customs union. In addition, in line with the EC's concessions on reference prices to its other Mediterranean Associates (outlined in the following section), from 1990 on there is the possibility of concessions on reference prices on sweet oranges, lemons and table grapes within specified quantitative limits.

Table grapes, an important export crop for Cyprus, can be taken as a case in point. Before the new arrangements came into force Cyprus enjoyed a preferential tariff for a limited period in the summer, but subject to a maximum quantity of 7,500 tonnes. This tariff quota remains, though the maximum quantity is to be progressively raised to 11,000 tonnes during the first phase of the second stage; and the tariff paid within the tariff quota is to be progressively reduced to zero. Reference prices will still have to be respected, though - as noted above - from 1990 on some concessions on the reference price system may be granted to Cyprus on a maximum quantity of 10,500 tonnes of table grapes within the same calendar limits.

It will be seen from the above that it is not the intention of this agreement to achieve free trade in agricultural products during the first phase of the second stage. Furthermore, Article 25 specifically provides that 'The application of the frontier mechanisms of the Common Agricultural Policy shall not be affected during the first phase of the second stage...' except for the limited concessions on reference prices to which reference has already been made.

Article 26 goes on to specify the prior conditions which must be met before implementing the second phase of the second stage, involving the free movement between Cyprus and the EC of the limited list of agricultural products covered by the Association Agreement and the Protocol (essentially fruit and vegetables and wine):

'i)   the introduction by Cyprus of Community quality standards for these products;

ii)   the application by Cyprus internally of domestic price constraints similar to those in force in the Community, with a view to ensuring the stability of the domestic market and avoiding market crises. In this connection and with a view to preventing recourse to safeguard measures, procedures shall be set up for the identification of a state of crisis on the market and provision for measures which Cyprus should apply on its domestic market in relation to the degree of disturbance or risk of disturbance;

iii)   the application by Cyprus of Community measures for these products at the Cyprus frontier.'

Thus, entry into force of the second phase of the second stage of the Cyprus: EEC Association Agreement will involve the adoption by Cyprus of the CAP, or CAP-compatible measures, for those products covered by the agreement. At the end of the second phase of the second stage 'The Customs Union shall be fully achieved' (Article 31). Customs Union clearly will not involve all agricultural products; the main items included are of export interest to Cyprus while those excluded are mainly imported into Cyprus. For those products included in the Customs Union, Cyprus will have to adopt CAP mechanisms; but the precise mechanisms to be adopted and the procedures to be followed have as yet to be determined.

REFERENCE PRICES AND PREFERENCES

The states of the Mediterranean Basin want, in the main, preferences on their agricultural exports to the EC. This has historically posed problems for the Community, for EC growers of Mediterranean products in Italy and Southern France have believed that their interests were sacrificed for the sake of the EC's wider Mediterranean diplomacy. Pomfret (1986, p 113) comments:

'The conflicts are becoming more severe because the EC is running out of room for manoeuvre. As self-sufficiency in CAP products rises, there is less scope for diverting imports to a preferred source. The Spanish accession will bring this to a head as Spain has the capability of filling (or overfilling) the gap between EC consumption and production of a number of products which are important exports of other Mediterranean countries ...'

The ability and willingness of Spain to increase its production, given the high opportunity cost of irrigation water, are perhaps less obvious than Pomfret implies; none the less a number of interesting policy issues arise.

First, it should be noted that the EC's system of preferences on agricultural products from the Mediterranean Basin have been rather more sophisticated, and more restrictive, than many observers have realised. For olive oil the general arrangement has been to allow a small reduction in the variable import levy (of 0.5 ECU/100kg) to allow the Mediterranean states a 'commercial' advantage on EC markets, and a second, but more substantial, levy reduction of 20 ECU/100kg conditional upon the Mediterranean states levying an equivalent export tax (Falgon, 1986, p 104). Paradoxically, in EC jargon, this second levy reduction is referred to as an 'economic' advantage even though it clearly does not allow EC market prices to be undercut. Moreover, the reference price systems for fruit and vegetables, and for wine, have had the effect of directly extending the EC's price support mechanism to the producers of its Mediterranean Associates.

The impact on consumers of the reference price system for fruit and vegetables has already been commented upon (Chapter 7). The system involves the monitoring of prices in EC wholesale markets, and if these fall below a minimum level - defined by the reference price - for produce from any particular origin, then subsequent shipments of that product from the 'offending' supplier country will face countervailing charges, until such time as the reference price is deemed to have been respected. (For further details see Ritson and Swinbank 1984, 1986).

Suppliers will no doubt strive to avoid paying countervailing charges. Thus it is very difficult to assess the combined protective effect of the CCT and reference prices: the fact that import duties on fruit and vegetables are relatively low is not evidence that the degree of protection is low. Alvensleben *et al.* (1986, pp 12 and 9), whilst recording that the average tariff burden on EC imports of fresh fruit and vegetables from Spain in 1982 was 11.2 per cent, also note that 'Spain has introduced a quota system for its exports to the EC' so as to control supply and avoid countervailing charges. Furthermore, given the volumes of citrus and most top fruit withdrawn from the EC market, it must be concluded that EC policy does act to support market prices (Ritson and Swinbank 1986, pp 45-49, Williams and Ritson 1987).

Reference prices for fruit and vegetables are expressed in ECU, and the conversion rate to and from national currencies has always been the market and not the 'green' rate. Thus, in terms of foreign currencies, the height of the protective barrier reflects the strength or weakness of that currency *vis-à-vis* the ECU. But, since the introduction of the 'green' ECU in March 1984, the protective barrier has in effect been increased by a further 14.5 per cent (see Table 15.2), for the correcting factor (that is, the 'green' ECU) which is used for MCA computations is also deployed in the calculations made to check whether or not reference prices have been respected.

Even where a tariff concession has been granted to the Mediterranean Associates, that concession - to date - has not allowed the preferential suppliers a price advantage on EC markets. The reference price system has acted to ensure that EC market prices are not undercut; instead the preference is reflected in a higher cif price for the preferential supplier, perhaps enabling some high-cost preferred suppliers to displace lower-cost non-preferred suppliers from the EC market.

In the revised agreements, Mediterranean Associates gain duty-free access on certain products, limited by tariff quotas based on traditional traded quantities, following a transitional period linked to the ten-year transition of Portugal and Spain; and some countries have been offered the prospect of an amendment to the reference price system from 1990, within quantity limits and for a limited number of specified crops, but again linked to the transitional arrangements in place for Portugal and Spain (see, for example, Articles 1 and 2 of the Additional Protocol to the Tunisian Co-operation Agreement; *OJ* L297, October 21, 1987).

The latter arrangements were first triggered in December 1989 when the Commission, under the Management Committee procedure, decided that for the citrus fruit concerned, although for somewhat smaller quantities than those specified in the Protocols, in calculating in 1990 whether or not the reference

Table 15.2: Coefficients for the 'Green' ECU

Date	Coefficient	Comment[*]
March 1984	1.033651	Introduction of system, reduction of 3 percentage points in German MCA
July 1985	1.035239	Devaluation of the Italian lira, and other changes within EMS
April 1986	1.083682	Revaluation of German mark and Dutch guilder, and other changes within EMS
August 1986	1.097805	Devaluation of the Irish punt
January 1987	1.125691	Revaluation of German mark, Dutch guilder and Belgian and Luxembourg francs
July 1987	1.137282	1987/88 CAP price fixing
January 1990**	1.145109	Devaluation of Italian lira

[*]   Each reshuffle within EMS involves the fixing of a new rate for sterling and the Greek drachma.
**   And all ECU support prices reduced by 0.17 per cent, effected by dividing all CAP support prices by the coefficient 1.001712.
Source: *Official Journal of the European Communities*, Commission (1987).

price had been respected, the amount to be deducted from the wholesale market price would be five-sixths of the full rate of duty (*OJ* L380, December 29, 1989). This was the maximum adjustment permitted, and it remains to be seen whether or not the Commission intends to follow this precedent year by year and increase the deduction to 100 per cent of the full rate of customs duty by 1995. These arrangements should allow some part - if not all - of the duty concession to be reflected in lower EC sales prices, at least on those quantities benefiting from the reference price concession (see Ritson and Swinbank 1986, pp 54-56, for further discussion).

Thus the Iberian Enlargement could involve both trade creation and trade diversion in the fruit and vegetables sector. The CAP, of course, is usually represented as the very antithesis of trade creation, for its mechanisms generally operate to ensure that high-cost producers are not displaced by lower-cost suppliers from elsewhere in the EC (Chapter 9). The concern of French and Italian growers, and Northern producers of hothouse produce, is that for once trade creation might prevail; but the danger is that CAP mechanisms will deflect lower-cost Spanish production into intervention.

For the purposes of the present chapter, the more relevant trade creation/trade diversion question relates to EC imports of fruit and vegetables from its Mediterranean Associates. In the main the access conditions for Spanish produce have been less generous than those afforded most of the other Mediterranean suppliers. Thus it is possible that higher-cost, but more preferred, producing countries which until now have supplied the EC will be displaced by Spanish competition. Displaced suppliers, whether victims of trade creation or trade diversion, might be thought to be indifferent between these theoretical niceties: but the genuine low-cost supplier, previously selling in the EC market over the full tariff barrier and now a victim of trade diversion, is in a better position to find alternative, non-EC, outlets than is its erstwhile competitor which was only able to maintain its EC market share by exploiting a preferred status.

Elsewhere (Ritson and Swinbank 1986, p 25, Swinbank 1987) we have suggested that the implications of EC Enlargement for the fruit and vegetables sector can be viewed in a step-by-step fashion: the further one progresses down the steps the worse the consequences for non-EC suppliers:

Step 1: Trade deflection (involving both trade diversion and trade creation). Spain captures a larger share of the more remunerative EC market, while ceding lower-priced non-EC markets to suppliers who have lost their EC market share;

Step 2: Spain does invest in irrigation facilities, and consequently its production expands, squeezing out imports from the EC market;

Step 3: EC surpluses emerge, but these are dealt with by the EC's withdrawal and intervention mechanisms;

Step 4: EC surpluses are dumped on world markets, using the export refund system.

Until stages 3 and 4 in the above progression are reached, the Mediterranean Associates still have some advantages to be gained from their preferential access arrangements, for the burden of the adjustment to the Iberian Enlargement can be shifted to non-preferred suppliers. The new arrangements for the reference price system for preferred suppliers, referred to above, would encourage such an outcome. Whilst the marginal supplies come from non-preferred suppliers, EC market prices will be influenced by the full reference price protective mechanism giving protection to both EC and (within quota) preferred suppliers. However, if supplies were more plentiful then EC market prices would weaken: EC and (within quota) preferred suppliers would still be able to compete on price, but non-preferred suppliers would face countervailing charges if they tried to do likewise. Similarly, within quotas, low-cost preferred suppliers could compete on price against higher-cost EC and preferred suppliers.

As yet it is far from clear how the new concessions on reference prices will work given that they are limited by quota. The previous method of checking respect of the reference price was to monitor wholesale market prices and deduct the full rate of CCT. The new concession involves the deduction of an abated rate of CCT (five-sixths of the full CCT for citrus in 1990). Presumably the abated rate of CCT will be used in the calculations until such time as the Commission declares the country quota full. For the preferred supplier to take full advantage of its position, the Mediterranean Associate will require even greater control of the marketing process because it might be necessary to engineer a significant increase in the wholesale market price at short notice.

All the Mediterranean states are keen to ensure that their renegotiated agreements maintain their preferential status. Indeed, a hierarchy of preferences is liable to generate divisive splits among the beneficiaries for, if, as a result of stages 1 and 2 of the above progression, it is not just non-preferred but also preferred suppliers which are squeezed out of the EC market, it matters to any one state how generous its preferences are in relation to its neighbours.

An expansion in Spanish production is not necessarily to be expected in terms of an increased output of current crops; rather the likelihood is that efforts will be made to extend the EC's protective mechanisms, particularly the reference price system, to prolong the season of traditional crops, and to permit the commercial production of new semitropical exotics. Producers in the South of Spain, for example, are reported to be investing in avocado orchards - until now an EC market developed and dominated by Israel.

CONCLUSIONS

The recent history of the EC's fruit and vegetables policy leads one to conclude that, unless thwarted by the GATT negotiations, the system is likely to become more, rather than less, protectionist; and in particular, as noted above, that non-EC growers which currently exploit a slight climatic advantage at the beginning or end of the EC's season, or produce semitropical products, are likely to see their market access severely curtailed. The EC's Mediterranean Associates must be expected to use such political influence as they have to ensure the maintenance of their toe-hold in the EC's reduced import market: Morocco's spurned application to join the EC (*Financial Times* October 3, 1987) might be better interpreted as a political stratagem in the Mediterranean states' jockeying for position in the revised hierarchy of preferences, than a serious bid for membership.

It would appear to be the EC's intention that the Mediterranean states should keep their preferences *vis-à-vis* less preferred suppliers. Indeed the Commission has suggested that 'limited changes could be made to quotas, ceilings and time schedules and that new products could be included to satisfy the requests of some Mediterranean partners' (Commission 1990, p 2). Thus, the countries most at risk from Enlargement and revision of the Mediterranean Agreements are the present, or *potential,* non-preferred suppliers of high-value, out-of-season produce. This pessimistic conclusion is reinforced by reports of major investment on irrigation schemes throughout the countries of the Mediterranean Basin, often funded by the World Bank, which have apparently been prompted by a myopic view of their export potential to the EC (Bale 1986, pp 2-3).

REFERENCES

Alvensleben, R. von, Behr, H.-C. and Jahn, H.-H. (1986) Fruits and Vegetables in the European Community. In: Bale, M.D. (ed) *Horticultural Trade of the Expanded European Community, Implications for Mediterranean Countries*, The World Bank, Washington DC.

Association Council (1987) European Economic Community and the Republic of Cyprus Joint Press Release, CEE-CY 702/87 (172), Brussels, October 19, 1987.

Bale, M.D. (1986) Overview. In: Bale M.D. (ed) *Horticultural Trade of the Expanded European Community, Implications for Mediterranean Countries*, The World Bank, Washington DC.

Commission of the European Communities (1987) *Report on the Agri-monetary System*, COM(87)64, CEC, Brussels.

Commission of the European Communities (1989) *The Agricultural Situation in the Community 1988 Report*, Office for Official Publications of the European Communities, Brussels and Luxembourg.

Commission of the European Communities (1990) *The Commission's Proposals for Redirecting Mediterranean Policy*, Information Memo P-34, CEC, Brussels, May 22.

Council of the European Communities, General Secretariat (1985) *Press Release 10707/85 (179)*, 1044th Meeting of the Council - Foreign Affairs - Brussels.

Council of the European Communities, General Secretariat (1986) *Press Release 9572/86 (148)*, 1110th Meeting of the Council - Research - Luxembourg.

Dam, K.W. (1970) *The GATT: Law and International Economic Organization*, University of Chicago Press, Chicago and London.

Economic and Social Committee (1990) Opinion on the Mediterranean Policy of the European Community, *Official Journal of the European Communities*, C168, July 10.

Falgon, C. (1986) The Effect of Enlargement of the EC on Tunisian Fruits and Vegetables. , In: Bale, M.D. (ed), *Horticultural Trade of the Expanded European Community, Implications for Mediterranean Countries*, The World Bank, Washington DC.

Gaines, D.B., Sawyer, W.C. and Sprinkle, R. (1981) EEC Mediterranean Policy and US Trade in Citrus, *Journal of World Trade Law*, 15(5), pp 431-439.

Genazzini, L. and Hörhager, A. (1987) Food Self-Sufficiency in the Southern Mediterranean Countries, *Cahiers BEI/EIB Papers,* European Investment Bank, September, pp 55-79.

GATT (General Agreement on Tariffs and Trade) (1985) *European Community - Tariff Treatment on Imports of Citrus Products from Certain Countries in the Mediterranean Region: Report of the Panel,* L/5776, GATT, Geneva.

Harris, S.A. (1977) *EEC Trade Relations with the USA in Agricultural Products,* Occasional Paper No 3, Centre for European Agricultural Studies, Wye.

Hine R.C. (1985) *The Political Economy of European Trade: an Introduction to the Trade Policies of the EEC,* Wheatsheaf, Brighton.

McGovern, E. (1986) *International Trade Regulation: GATT, the United States and the European Community,* 2nd edition, Globefield, Exeter.

Pomfret, R. (1986) The Trade-Diverting Bias of Preferential Trading Arrangements, *Journal of Common Market Studies,* 25(2), pp 109-117.

Ritson, C. and Swinbank, A (1984) Impact of Reference Prices on the Marketing of Fruit and Vegetables. In: Thomson, K.J. and Warren, R.M. (eds) *Price and Market Policies in European Agriculture,* University of Newcastle upon Tyne.

Ritson, C. and Swinbank, A. (1986) *EEC Fruit and Vegetables Policy in an International Context,* Agra-Europe Special Report No 32, Agra-Europe, Tunbridge Wells.

Swinbank, A. (1987) *EEC Enlargement and the Question of Market Access,* CAP Briefing No 6, Catholic Institute for International Relations, London.

Swinbank, A. and Ritson, C. (1988) The Common Agricultural Policy, Customs Union and the Mediterranean Basin, *Journal of Common Market Studies,* Vol. 27, No. 2, pp 97-112.

Tsardanidis, C. (1984) The EC-Cyprus Association Agreement: Ten Years of a Troubled Relationship, 1973-1983, *Journal of Common Market Studies,* 22(4), pp 351-376.

Williams, H. E. and Ritson, C. (1987) *The Impact of the EEC's Reference Price System on the Marketing of Fruit and Vegetables in the UK,* Report No 31, Department of Agricultural and Food Marketing, University of Newcastle upon Tyne.

PART IV

THE CAP AND THE FUTURE

CHAPTER 16

THE REFORM OF THE CAP

Lionel Hubbard and Christopher Ritson

INTRODUCTION

Virtually since its inception, the Common Agricultural Policy has been subject to proposals for reform. Among the academic community anyway, the debate rapidly achieved something of a consensus and subsequently has evolved remarkably little. Academics and other CAP specialists, either as individuals or in groups, have produced numerous reform proposals. Notable in this context have been the Wageningen Memorandum of 1973 and the Sienna Memorandum of 1984 (Ritson 1984) - because of the number of academics from various European Community member states who were willing to add their signatures to the documents. Reading these documents one is struck by how little seemed to have changed - even some of the people are the same! A typical form of this kind of argument is outlined in the next section.

THE ACADEMIC REFORM ARGUMENT

The argument would begin as follows. When the Agricultural Ministers of the original six Common Market countries launched the CAP, they made a fatal mistake. Partly because of a failure to appreciate the potential growth in production, and partly because of the overriding necessity to achieve a political agreement in agriculture to cement the establishment of the European Economic Community, they set support prices for cereals at too high a level (more than 50 per cent in excess of, what was then, a very stable world market price). As a consequence, because most other agricultural products are related to cereals, either as competitive arable crops or as users of cereal-based feeding stuffs, most other agricultural product prices had similarly to be set at relatively high levels.

From this single set of decisions there followed, so it was argued, a number of undesirable consequences - undesirable, that is, when viewed against a set of criteria, widely accepted (though often implicitly) among academics, concerning agricultural policies. This approach judges the success of agricultural policies relative to certain fundamental goals in society, such as the efficient use of resources and equity in income distribution (in each case, both within agriculture and between agriculture and other sectors) and good international relations. Because support prices were so high, it was argued that the CAP encouraged inefficient, high-cost production; impeded structural adjustment; disadvantaged low-income consumers (because of high food prices); benefited large farmers greatly and small farmers very little (because the benefit was distributed pro rata to the amount produced); and damaged trading relations with both rich and poor countries alike. All would be well, however, if agriculture product support prices were reduced - and the debate, as such, was really about how to achieve this.

The problem, of course, was the damage to farm incomes that would result from lower prices. The key, in most reform proposals, was the introduction of some form of direct income supplementation for low-income farmers. Such a switch, from price support to direct income support for agriculture, seemed highly desirable when judged by the criteria listed above. Greater equity could be achieved, on account of the benefit to low-income consumers of lower food prices and the ability to target support to those farmers most in need; efficiency could be improved, in that the Policy would no longer underwrite high cost marginal output; and international relations and the prospects for developing country exports of one or two key products would be improved. Such a development seemed to favour the prospects of the low-income country exporters to the EC, and so the cause of reform of the CAP has typically been embraced by the development lobby - though, as pointed out in Chapters 12 and 13, the issue is far from clear-cut.

THE FORMAL APPROACH

Most of the literature on CAP reform, although written by academics, was directed towards a more general audience with the aim of influencing policy-makers. Underpinning it, however, was other more professional analytical work which, with the benefit of hindsight, can be seen to originate with the publication in 1969 of Josling's article 'A Formal Approach to Agricultural Policy'. It is a peculiar feature of post-war agricultural economics that, whereas an analytical approach to farm production economics was well advanced by 1969, agricultural policy had tended to be the preserve of a more 'literary' type of agricultural economist. Josling applied public choice theory to agricultural policy - attempting to analyse the appropriate choice of policy in terms of objectives, constraints and instruments.

In Figure 16.1 we illustrate, first, the idea of instruments and complementary objectives - that is where the instruments all have a positive relationship with the objectives - which are here, by way of example, to raise farm output (perhaps for balance of payments reasons) and to raise farm incomes. For the sake of argument, the instruments might be product subsidies, import controls and investment subsidies - and instrument 3, to illustrate the various possibilities, is drawn so that, past a certain level, the objectives are no longer complementary.

Figure 16.1:   Policy Instruments and Objectives

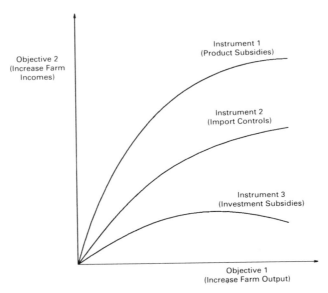

Now the important message from the diagram, and to quote the Josling (1969) article, is that: '...a necessary (though not sufficient) condition for the reaching of a number of quantitative objectives is that one employs a similar number of policy instruments'. In other words it is extremely unlikely that the choice of any one of the instruments in Figure 16.1 would allow the achievement of target levels of both objectives.

Thus, the more 'academic' version of the argument conveyed in the previous section was that 'the CAP is unsuccessful because it is attempting to use one instrument (raising market prices) in order to fulfil a number of objectives'.

OFFICIAL REFORM PLANS

Meanwhile, it was not just academics who were calling for the reform of the CAP - but also successive European Commissions. Indeed, perhaps the most famous reform plan of all - the Mansholt Plan - came from the Commission of the EC (1968), and was in general consistent with (and indeed largely pre-dated) the academic work. It advocated reduced prices for surplus commodities, but linked various forms of financial compensation to structural reform. It was either 'get out or get bigger', whereas other reform proposals were willing to contemplate 'stay small - here is a bit extra to supplement your income from farming'. [1]

The Mansholt Plan was far from popular with the agricultural interests in Western Europe, but subsequently there have been a succession of Commission documents proposing change to the CAP - though after Mansholt the term 'reform' became *infra dig* - euphemisms such as 'adjustment', 'development', 'guidelines' and 'improvements' have been used. [2] Gradually, however, as we moved into the 1980s, the preoccupations of the European Commission, and some other CAP commentators, diverged from those of what one might call 'academic orthodoxy' - and a parallel debate developed. This is now the real debate, in the sense that its preoccupations are those which are driving change.

For an academic, it is somewhat sobering to realise that the arguments involved in the consensus over CAP reform mentioned earlier are almost wholly irrelevant to the actual reform of the CAP. Herein lies a paradox, for it was the peculiarity of having to forge a common agricultural policy for six countries which was partly responsible for the original mistakes with respect to the CAP; and it has been the fact that the Policy has had to meet six (and then nine, and now twelve) member state interests which has partly been responsible for the failure of the Policy to reform in the way advocated by many; but it is another peculiarity of the Common Policy which has driven change. This is the failure of the Community's automatic system of generating revenue (its 'own resources') to match the budgetary cost of its policies - mainly the cost of the agricultural policy.

REAL CAP REFORM AND PUBLIC CHOICE THEORY

Does this mean that the theory of public choice is of no relevance in practice to understanding the development of the CAP? We think not; rather it has been a failure of the academic work to catch up with the changed policy environment.

First, the 1980s have seen a proliferation of policy instruments under the CAP, which can be viewed as coming to terms with the need for more than one instrument to attain more than one objective. Second, the priority of objectives and the relation between objectives and constraints are largely a political issue. Figure 16.2 considers the second case, where the instruments are associated with conflict between the objectives. We have chosen to insert a new objective 'economic efficiency', but could also have introduced a new instrument (say quotas) to Figure 16.2. Product subsidies and import controls both adversely affect efficiency (but import controls more so, because they raise consumer prices).

---

1   For a full discussion of the Mansholt Plan, see Marsh and Ritson (1971).

2   The 'evolution' of Commission reform proposals is traced in a paper (Ritson and Fearne 1984) prepared for the Sienna meeting, which led to the 'Sienna Memorandum' as mentioned earlier.

Figure 16.2: Policy Objectives : Trade-offs

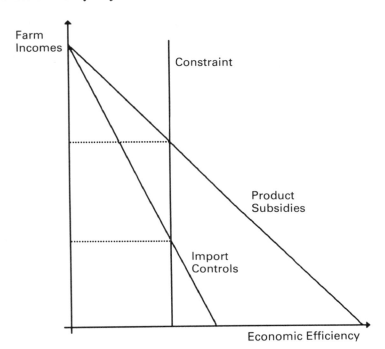

In these circumstances choosing an optimum policy is best seen as maximising one objective subject to viewing the other objective as a constraint. One then chooses the instrument which achieves the highest value of the objective (say farm incomes) subject to the constraint (efficiency). In this example product subsidies would be chosen.

In the 1970s farm incomes were usually taken as the prime objective, with equity, import saving and efficiency as subsidiary objectives. Public expenditure was seen, if at all, as a constraint. What seems to have happened to the CAP is that what was once best viewed as a constraint is now a major objective. Why this should have become so is now discussed.

THE BUDGET PROBLEM

During the past fifteen years the European Community moved from net importer to a net exporter in the case of several major agricultural commodities (see Figures 16.3-16.6). To the academic CAP specialist, at the time, the point at which the EC moves from 100 per cent to, say, 105 per cent self-sufficiency in a commodity was seen as of no greater significance than when it moves from 95 per cent to 100 per cent. High-cost, inefficient output is still just that, whether or not the cost is expressed as an export subsidy or as a levy which displaces cheaper imported produce; high prices are still high prices to consumers and producers; and the world trading system is affected in much the same way by the elimination of a previous import requirement, as it is by the addition of a new export availability. But, from the point of view of the administration of the Common Agricultural Policy, the movement into export surplus matters a great deal.

This is because it represents the point at which the Community budget begins to have to 'top up' the contribution of food consumers to enhancing the revenue of farming. When the Community is less than 100 per cent self-sufficient, market prices can be supported solely by taxing imports. When EC production exceeds EC consumption, however, prices can only continue to be supported if the Community finances the purchase and disposal of surplus production; and, as surpluses grow, so does the budgetary cost of their disposal. By the early 1980s, the cost of surplus disposal had taken EC expenditure to the ceiling imposed by 'own resources', and the Community was forced to resort to various devices to balance the books. The most significant of these was to allow stocks to accumulate.

Member states bear the cost of intervention until stocks are disposed of. By manipulating the level of export refunds, the Commission is able to control the timing of exports and effectively transfer intervention expenditure forward into the next financial year. Stocks have also accumulated for two further reasons. First, in the case of dairy products, there are very few export outlets even for subsidised produce (a consignment delivered to Russia went at 13 per cent of the intervention price). Second, in the case of cereals, the fear of retaliation by the United States has limited somewhat the willingness of the Community to subsidise exports to the traditional markets for North American produce. But accumulation of stocks only delays the time at which the EC budget must bear the cost of surpluses. The consequence is that, to understand the reform of the CAP, it is necessary to appreciate that there is really only one criterion at work, epitomised by the question 'Does the change make a positive contribution to the Community budget?'; and, in attempting to classify the various policy changes under way or under discussion, it is most helpful to categorise them according to the main way in which they might be expected to contribute positively to the EC budget. This is done in Table 16.1.

In choosing between alternative policies which have similar budgetary implications, a second criterion will be apparent. This is that the reform will be favoured which involves least change in the current balance of benefit for the interest groups affected by the CAP - particularly when one interest group is biased towards some member states - which is nearly always the case.

The reform measures listed in Table 16.1 are categorised according to budgetary effect. Broadly, there are two groupings - those which seek to reduce expenditure and those which seek to increase revenue. These will be dealt with in turn.

Figure 16.3: EC Cereals Balance

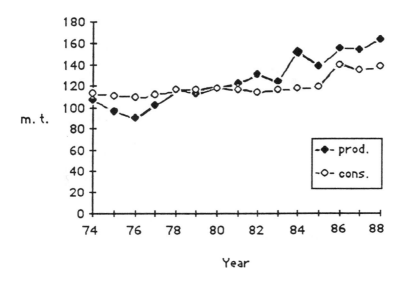

Figure 16.4: EC Sugar Balance

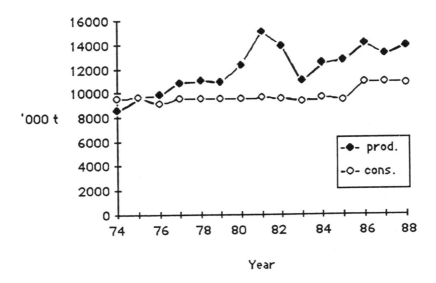

Figure 16.5: EC Butter Balance

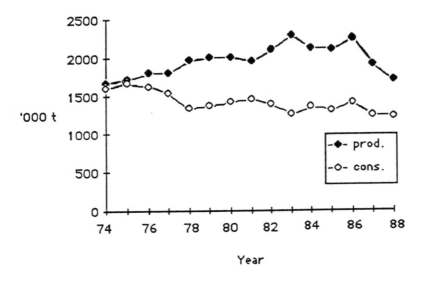

Figure 16.6: EC Beef and Veal Balance

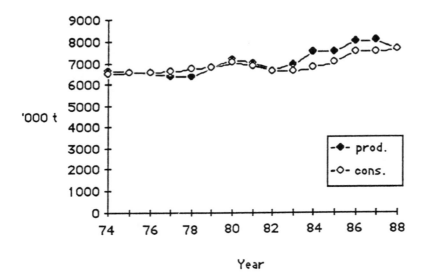

Table 16.1: CAP Reform Classified According to Budgetary Effect

---

Expenditure-Reducing
 1) reduce support prices
   a) price cuts
   b) stabilisers
   c) intervention criteria
 2) limit production
   a) marketing quotas
   b) production quotas
   c) 'set-aside'
Revenue-Increasing
 1) product taxes
   a) co-responsibility levies
   b) vegetable oil tax
 2) input taxes
 3) national financing of CAP policies
 4) new sources of revenue
   a) related to agricultural production
   b) related to GNP

EXPENDITURE-REDUCING MEASURES

There are two dimensions to this particular objective: a) the quantity of surplus produce, and b) the unit cost of disposal (essentially the gap between EC intervention prices and international trading or 'world' prices). The first possible reform measure cited is that of reducing support prices. This is the measure favoured most by 'market economists'. Lowering the intervention price for a commodity will secure an immediate reduction in the unit cost of disposal. Subsequently, it should reduce (or at least impede the growth in) production and therefore the size of any surplus. (This assumes, fairly rationally, that the supply of a commodity is positively related to its price.) In the longer term such a policy measure may be expected to reduce yet further the unit cost of disposal, with the world price rising as a consequence of reduced EC exports. It has been argued by the Australian Bureau of Agricultural Economics [3] (1985) that the CAP depresses world prices by between 9 and 28 per cent.

Experience suggests, however, that although in recent years support prices have been 'frozen' (in common ECU terms) and have fallen a little in real terms in recent times (see Figure 16.7), and, although the Commission will continue to put pressure on the Council of Ministers to moderate prices, it is simply not

---

3 Since renamed the Australian Bureau of Agricultural and Resource Economics.

possible from a political point of view to cut prices sufficiently to constrain the cost of the CAP. Whilst straightforward price cuts are the economist's solution to the problem, they are widely accepted as being politically all but impossible, as evidenced by the extreme difficulties that the Agricultural Council had in 1990 in agreeing even the modest price cuts implied by the Commission's proposal to reduce the 'Aggregate Level of Support' by 30 per cent over a ten-year period, as the EC's offer under the Uruguay round.

Since straightforward price-cutting has proved so difficult a task for the Council of Ministers, price reductions are now embodied in the agricultural 'stabilisers'. These exist for the major commodities and are designed to stabilise markets. Each stabilisation mechanism is tailored to the specific features of the relevant product, but the basic idea is always the same: whenever production breaks through a ceiling set in advance, support is automatically reduced for that product. (Commission of the European Communities, 1987). The production ceiling for a commodity is referred to as the 'maximum guaranteed quantity', and when this is exceeded a price reduction follows automatically in the subsequent season. In principle, the system has an advantage over earlier attempts at price-cutting, in that its effects are automatic and do not require approval from the Council each year. In practice, its major shortcoming is that, whilst price cuts are automatic, there is no such restraint on the setting of 'gross' prices. Thus, the price cut relates to a price which still requires the agreement of the Council of Ministers. If there is to be any significant impact on curtailing supply, these 'gross' prices will need to be kept in check. Thus, to date, the system of price determination is still open-ended. Nevertheless, there were some large reductions in prices for certain commodities in the late 1980s as a result of the stabiliser mechanism (for example, oilseeds), and the recent agreement on a 30 per cent cut in support to agriculture may signal a move to lower levels of 'gross' prices.

Over recent years a number of changes have been made in the rules governing intervention buying. Minimum quality standards have been raised, the times of the year when intervention buying is undertaken have been shortened, and the prices at which commodities are purchased by intervention agencies have been reduced, that is, the 'buying-in' price and the intervention price have become separated. In bringing about these changes the Commission has sought to persuade farmers to produce for market rather than for intervention, and to restore the intervention agencies to their original role of 'buyer of last resort'. With intervention now a less attractive option, further budgetary savings have been created.

The second group of expenditure-reducing measures is that which physically limits production. Viewed from the narrow criteria of budgetary effects and damage limitation to the existing balance of interests, quotas have considerable attractions. Production can be reduced very quickly to whatever level is regarded as manageable, and since prices do not need to be reduced as well (indeed they are likely to be increased in compensation, for most producers) the impact on net income is likely to be only marginal. In the case of milk quotas, introduced in great haste in April 1984, production has been reduced significantly. Deliveries of milk to dairies in 1988 were some 10 per cent below the peak reached in 1983, immediately prior to the introduction of quotas. Butter production was down by 26 per cent over the same period, and production of skim-milk powder down by 47 per cent. As a consequence, intervention stocks of these dairy products have fallen, although, with milk production in the Community still exceeding domestic consumption, the sector remains the most expensive within FEOGA. Quotas create innumerable problems and inequity but, perhaps somewhat surprisingly, have been integrated into Community agriculture more easily than had been originally envisaged. In part this is because, faced with a choice between quotas or a price cut equivalent, most producers will be better off opting for quotas. But there are administration problems for the authorities.

Figure 16.7: Gross FEOGA Expenditure and Annual Price Changes

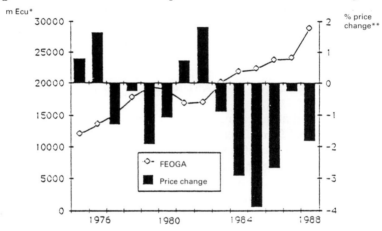

* Gross FEOGA expenditure in 1988 values.
** Annual average real price change in national currencies.
Source: Commission of the European Communities (1968).

Marketing quotas (essentially, where a limit is placed on the quantity of output from each producer eligible for price support - as with the arrangements that now apply to milk) are only feasible where the major part of production has to pass through a number of controllable processing units. This is possible with milk (and sugar), but more difficult with many other products, in particular cereals. Thus, there have appeared proposals for restricting what is produced, rather than what is marketed - though of course the response to a marketing quota is probably to reduce production. The most obvious method of directly limiting production is to control the amount of land that can be devoted to certain crops - or 'set-aside' as it is commonly known.

305

Experience of set-aside in the United States, where it has been employed intermittently since the 1930s, is not encouraging. For a number of reasons, the impact on production is not as great as might be imagined, and any reduction that does occur tends only to be short-lived, as advancing technology continues to improve yields. Indeed, this last point is particularly pertinent in the case of the EC, where a voluntary set-aside policy for cereals has now been introduced, since production over the last 20 years has risen almost entirely as a result of better yields from virtually the same land area. Work undertaken to ascertain the impact of set-aside suggests, therefore, that it needs to be combined with longer-term policies, such as price reductions. [4] Set-aside has many critics, and from the economist's viewpoint is perhaps most appropriately seen as an attempt on the part of the authorities to meddle with one of the inputs, albeit a rather special one, in the production process. In this respect, set-aside has some similarities with input taxes (see below). From a purely budgetary angle, the important question to be answered is whether the cost of the set-aside policy, in terms of the payments that have to be offered to farmers to induce them to idle part of their land, will be less than the cost of surplus grain disposal. If the former outweighs the latter, as some critics claim, set-aside offers no solution to the budgetary problem.

REVENUE-INCREASING MEASURES

Policy measures aimed at increasing the Community's budgetary revenue can seek to raise additional funds from various sources. With product taxes the intention may be to shift part of the financial burden of surplus disposal on to producers. This is the idea behind the co-responsibility levy, a measure based on the so-called 'self-financing' production levies operated in the sugar sector. Introduced for milk in 1977 and for cereals in 1986, the co-responsibility levy is a flat-rate tax on all (or most) of a product passing through marketing channels. Essentially, it means that part of what the consumer pays for the product is diverted from its previous path to the producer, and into the Community budget. Whether the levy acts also as a producer price cut (and thus discourages production) depends on what the ministers do to support prices. Provided the levy is high enough, it is quite possible to make a particular CAP commodity regime 'self-financing', with the tax revenue used to finance surplus disposal. This is the direction in which the sugar regime has gone; the milk and cereal co-responsibility levies are very small in comparison.

Co-responsibility has long been a favourite with the Commission and in fact dates back to the earliest days of the EC. The principle - to make producers bear at least part of the cost of surplus disposal - seems eminently sensible. In practice, co-responsibility has had a chequered history. The main problem is

---

4   See, for example, MAFF (1986), Buckwell (1986) and Hope and Lingard (1988)

essentially the same as that outlined above for the 'stabilisers', in that, whilst the co-responsibility levy reduces the price received by the producer, no formal restraint operates on the setting of the 'gross' support price by the Council of Ministers. Thus, it is not difficult for the levy to be transformed into a consumer tax, thereby avoiding any real hardship to producers (Hubbard 1986). Even so, it remains a fairly unpopular measure with farmers.

Included in the 1987 annual policy package from the Commission was a proposal for a tax on vegetable oils (strictly speaking, a 'consumer price stabilisation mechanism') which would operate in a similar way to a co-responsibility levy. The difference with oilseeds is that, because import tariffs are bound in GATT, EC support has taken the form of deficiency payments, and the oilseeds policy has become increasingly expensive. Domestic production has expanded and low market prices have required substantial budgetary payments. The effect of the tax would be to reduce the deficiency payments on domestically grown oils, and to raise revenue from oils based on imported oilseeds. The Commission claims that consumption and imports would not be affected - this is doubtful. In the event, the proposal was not accepted, which was probably just as well if American threats of a trade war were to be taken seriously. However, the proposal may surface again at some future date.

An alternative measure of raising revenue, and one which has the added attraction of reducing production at the same time, is that of an input tax. This will raise the price of an input, and discourage producers from using it. Recently, interest has centred on the possible use of a tax on nitrogen fertiliser. This has yet another attraction - that of lowering what are becoming dangerously high levels of nitrate in ground water. Rickard (1986) has argued that reduced nitrogen usage would inflict least harm on the smaller farms - a feature which the Commission may find particularly appealing. However, there is considerable debate about the extent to which production is really sensitive to fertiliser use and, more particularly, the extent to which the fertiliser use would be sensitive to rises in its price. So whether such a tax would have a marked impact on production is unknown but it would, of course, raise revenue. Its main attraction lies in the unusual coincidence of the preoccupation of CAP reform, as discussed here, and environmental concern.

Another area of reform, which, in part, is already under way, is that of national financing of the CAP. Member states have been responsible for partial financing of some regimes - for example, butter disposal measures and various production premiums in the beef sector. However, the main area where a major degree of national financing could prove significant is if price cuts are eventually shown to be the only way of meeting the budgetary objective (or prove unavoidable if the world trading system is not to collapse because of a failure to reach agreement for agricultural products under the GATT Uruguay round). In this case farmers may have to be compensated by national governments (*Agra-Europe* 1987a). The main concern here is that a move to national financing of agricultural policies on a significant level would be seen as a regressive step and a departure from the principle of the single market. A similar but more acceptable development might be to link the financing of surplus production to those member states judged (by some criteria) to be most responsible, or to base member states' budgetary contributions on their shares of total Community agricultural production, rather than on value added tax and GNP, as at present (Buckwell *et al.* 1981).

Traditionally, the EC has obtained revenue from its 'own resources', comprising customs duties, agricultural levies and the VAT-based contributions of member states. Recently, a fourth own resource, related to member states' GNP, has been added to these original three. Whilst GNP-based contributions will alter the distribution of net costs (costs minus benefits) between member states on to a more equitable basis, it is unlikely to make it any easier in the future for the Commission to increase the total amount of revenue available. In fact, the 'budgetary discipline' now imposed by the Council of Ministers restricts the annual rate of growth in the Guarantee Section of FEOGA to 74 per cent of the annual growth rate of the Community's GNP. Whilst this has put a ceiling on the level of agricultural expenditure for any given year, it is unlikely to bring to an end the quest to reduce the budgetary burden and the need for further reform of the CAP.

THE FUTURE OF REFORM

This chapter has traced the development of the CAP reform debate from its origins within the academic community to a discussion of the forces which have motivated reform during the 1980s and into the 1990s. The reform of CAP in practice has been characterised, in the academic language of formal analysis of agricultural policies, as 'what was once best viewed as a constraint [finance] is now a major objective'. There seems no reason to believe that this interpretation of CAP reform will not remain valid as we move towards the year 2000.

Newcastle University has published a report on a Delphi survey of the future of the Common Agricultural Policy (Fearne and Ritson 1987) carried out among about 100 CAP experts throughout the Community. Among the conclusions of the experts concerning the 'CAP in 1995' were:

a) The CAP will remain (albeit somewhat more nationalised) in a Community of Twelve until 1995.

b) The Policy will, however, become increasingly dominated by budgetary pressures, which will result in stricter limitations of intervention and an increase in the general application of producer co-responsibility.

c) Budgetary resources will be increased, but the CAP (Guarantee Section) will continue to take the lion's share.

d) Average farm incomes across the Community will remain at current levels, although disparities between member states will remain.

e) The level of price support over the next 10 years will fall (in real terms) by around 10 per cent.

f) Dairy quotas are expected to remain beyond 1990, with the likelihood of further reductions in member state allocations.

Table 16.2 lists the responses to the question 'In which way is the method of support under the CAP likely to change by 1995?'

Overall, perhaps the most important message of the survey is to counteract the sensational 'CAP on the brink of collapse' view, which so often appears in the media. To characterise the contents of the Newcastle Report (as did *Agra-Europe* 1987b) as 'experts foresee no change CAP in 1995' is an over-reaction to the contrast with what normally appears in the press when reporting the future prospects for the CAP. But the considered view of experts throughout the Community does seem to be that the Policy will continue to bump along from one crisis to the next, adapting slowly and in a piecemeal fashion. There will be no dramatic radical reform; nor will the Policy 'collapse'.

Table 16.2: Future of the CAP

Policy Change	% of Responses
Limited Intervention	17.9
Extended Co-responsibility	16.1
Direct Income Aids	15.6
Structural Aids	13.8
Extended Guarantee Thresholds	11.6
Less Border Protection	8.1
Set-Aside Schemes	7.3
More Quotas	5.6
Other	3.9

Source: Fearne and Ritson (1987).

REFERENCES

Agra-Europe No. 1225, March (1987a).

Agra-Europe No. 1238, June (1987b).

Australian Bureau of Agricultural Economics (1985) *Agricultural Policies in the European Community - Their Origins, Nature and Effects on Production and Trade,* Policy Monograph No. 2, Canberra.

Buckwell, A.E. (1986) *Cereals Set-aside in the European Community,* Paper presented at the Agricultural Economics Society Conference, Reading University, October 1986.

Buckwell, A.E., Harvey, D.R., Parton, K.A. and Thomson, K.J. (1981) Some Development Options for the Common Agricultural Policy, *Journal of Agricultural Economics,* Vol.32 No.3.

Commission of the European Communities (1968) *Memorandum on the Reform of Agriculture in the European Community* (the Mansholt Plan), COM (68) 1000, Brussels.

Commission of the European Communities (1987) *Implementation of Agricultural Stabilizers,* Vol. 1, COM (87) 452, Brussels.

Fearne, A. and Ritson, C. (1987) *The CAP in 1995,* Report No. 30, Department of Agricultural Economics and Food Marketing, University of Newcastle upon Tyne.

Hope, J. and Lingard, J. (1988) *Set-aside - a Linear Programming Analysis of its Farm Level Effects,* DP8/88, Department of Agricultural Economics and Food Marketing, University of Newcastle upon Tyne.

Hubbard, L.J. (1986) The Co-responsibility Levy - A Misnomer? *Food Policy,* Vol. 11, No. 3.

Josling, T. (1969) A Formal Approach to Agricultural Policy, *Journal of Agricultural Economics,* Vol. 20, No. 2, pp 175-191.

Marsh, J. and Ritson, C. (1971) *Agricultural Policy and the Common Market,* PEP, London

MAFF (Ministry of Agriculture, Fisheries and Food) (1986) *Diverting Land from Cereals (Note by the UK),* Paper presented at the Agricultural Economics Society Conference, Reading University, October 1986.

Rickard S., (1986) *Nitrogen Limitations: a Way Forward?,* Paper presented at Agricultural Economics Society Conference, Reading University.

Ritson, C. (Rapporteur) (1984) Sienna Memorandum, The Reform of the Common Agricultural Policy, *European Review of Agricultural Economics,* Vol. 11, No. 2.

Ritson, C. and Fearne, A. (1984) Long Term Goals for the CAP, *European Review of Agricultural Economics,* Vol. 11, No. 2.

Wageningen Memorandum (1973) Reform of the European Community's Common Agricultural Policy, *European Review of Agricultural Economics,* Vol. 1, No. 2.

CHAPTER 17

THE PRODUCTION ENTITLEMENT GUARANTEE (PEG) OPTION [1]

David Harvey

PREFACE

This concluding essay is deliberately prescriptive, and its language is accordingly emotional and provocative. To echo Professor Ashton, [2] 'prices and markets provide important signals to producers and consumers with regard to the efficient allocation of resources within agriculture and to purchasing by consumers, yet our diverse systems of intervention commonly lead to substantial misallocation of resources'. It is, however, often insufficient to point out the costs of departing from the market mechanism as an allocative device. It is sometimes necessary to present alternatives which might be able to meet political, social and economic needs rather better than the existing set of policies. In so doing, it is not enough to present the traditional economists' alternative - the free market - without recognising that there are strong, not to say always legitimate, reasons for intervention and support in the domestic agricultural sector. This recognition forces consideration of policies which seek to maintain support while removing the worst, if not all, the misallocation consequences. While dispassionate argument and analysis have their place in such a presentation, the fact that the alternative is being proposed as a genuine, practical and acceptable alternative (cf. the free market alternative) means that some passion in the presentation is both in order and possibly a requirement. In any event, it is a genuine reflection of the author's feelings on the subject.

---

1   Lionel Hubbard (Newcastle University), David Blanford and Harry de Gorter (respectively Professor and Assistant Professor of Agricultural Economics at Cornell University, Ithaca, New York), have also been involved in the development of this proposal (see IATRC 1988 and Hubbard and Harvey 1988). Helpful editorial comment from Dr J. Lingard is also gratefully acknowledged.
2   Presidential Address to the European Association of Agricultural Economists, reprinted as Chapter 2 of this collection.

INTRODUCTION

World agricultural trade is in a critical state. Despite temporary reprieves, occasioned lately by the American drought, the condition is terminal. World grain prices are again under pressure as supplies return to their 'normal' level. The moral obscenity of food mountains in the developed world, contrasting starkly with shortages and famines in the developing world, is compounded by the economic insanity of developed country competitive subsidisation. [3] The European Community and the United States seem bent on outspending each other to retain or conquer world markets, with the result that domestic support policies become ever more expensive and world market prices ever more depressed. Forty-five per cent of the world's population is dependent on agriculture and lives in the developing world, with incomes only one-fifth of the minimal prosperity levels in the lesser developed countries. Depressed world conditions and lack of access to high-income markets are critical to their survival. Action which precipitates these conditions is little short of criminal.

In the European Community and the United States, farm policies are unsustainable. There is neither sufficient money in exchequers nor capacity in well-fed stomachs to cope with the growing surpluses. There are three options: cut support prices and market prices until domestic markets balance; control production through such instruments as quotas and acreage set-asides (the paid idling of land); or subsidise (dump) exports of surpluses on already saturated world markets. Combinations of all three are being used on both sides of the Atlantic in a desperate attempt to sustain domestic policies and the farm incomes which they are designed to support. While more than half the world's population is short of food, a large part of the developed world is curtailing supplies, while also dumping surpluses at the expense of world prices and production incentives, particularly for the developing world.

It is against this background that the latest round of international trade negotiations (the Uruguay round) under the General Agreement on Tariffs and Trade (GATT) had agricultural trade at the top of its agenda for the first time. The primary reason for agriculture's inclusion, however, is the unsustainable state of developed country agricultural policies. A major objective of the current Uruguay round is to eliminate trade distortions created by domestic agricultural policies so as to improve world markets and so reduce the necessity for domestic support.

Several proposals were put forward during the round, including one by the United States for the elimination of all (trade-distorting) subsidies in ten years. The European Community and Japan were reluctant to embrace such a commitment to complete elimination, which they fear is a commitment to eliminate support for domestic agriculture. Until the last moment, there was a stand-off between the European Community and the United States over this

---

3   The subject is dealt with, *inter alia,* in Harvey (1988).

issue, with little sign of compromise. The Ministerial 'mid-term' review of the negotiations, in Montreal in December 1988, achieved little more than a consensus that 'agricultural policies should be more responsive to international market signals in order to meet the objective of liberalisation of international trade and that support and protection should be progressively reduced and provided in a less trade distorting manner'. [4]

The basic options left for decision for the April 1989 meeting of the negotiating group were: a) 'whether the ultimate goal should be the elimination or substantial reduction of trade-distorting support and protection'; or b) 'whether this reduction or elimination should be realised through negotiations on specific policies and measures or through the negotiation of commitments on an aggregate measurement of support, the terms of which would have to be negotiated - or through a combination of these approaches'. Ministers are also invited to agree an overall freeze on existing support levels and measures and an x per cent reduction in these levels by 1990. By the autumn of 1990, the EC were talking about a 30 per cent reduction in support (basis 1986 levels of support), though still arguing about whether export refunds (subsidies) should be given special status in this commitment, while the US was proposing a 70 per cent cut from present (1989) levels, in which cuts in export subsidies would be given priority. As this book goes to press, the outcome is still unknown.

The consequences of failure to agree on substantial reductions, if not elimination, of trade distortions in agricultural trade are dreadful. For the developed world, continuation of current policies condemns governments to continual agricultural trade conflicts with potential spillover into other areas and escalation into trade wars. Furthermore, it condemns governments to continue and increase highly inefficient support expenditures and mechanisms, which simply offset the consequences of trading partners' policies without improving the domestic situation. More important still, failure to liberate world agricultural trade condemns three-quarters of the world population to perpetual misery with no prospect for improving their own circumstances.

The difficulty of obtaining an agreement in agriculture now, and in previous rounds of the GATT, can be traced to a preoccupation amongst some of the participants in the negotiations with reducing or eliminating farm income support, rather than reducing or eliminating trade distortions. Income protection is the primary objective of domestic agricultural policy in most industrial countries. This will not be given up easily or completely. Trade distortions, however, are an unwanted by-product of domestic protection, and successive multinational declarations [5] demonstrate that many countries

---

4   GATT (1988), pp 10-11.

5   The Punta del Este declaration launching the Uruguay round in 1986, subsequently reinforced at the 1987 OECD Ministerial Council and the 1987 Economic Summit.

are keen to reduce and eventually eliminate these. [6] While it can be hoped, as this is written, that the debate will be rendered redundant by a radical and far-reaching conclusion to the GATT Uruguay round, the prospects do not look good.

The critical question is whether there are policy instruments which allow national governments to continue support for farm income while reducing or eliminating international trade distortions. In order to reduce the effect of government policies on trade, economists have advocated 'decoupled' income support (direct payments unrelated to level of production). It is difficult to identify such distortion-free transfers in practice except for unanticipated 'lump sum' transfers. Although such decoupled options have desirable characteristics in theory and should be encouraged, several major shortcomings can be identified.

First, fully decoupled income supports are likely to be unattractive to policy-makers and farmers since they represent a radical departure from traditional commodity-based support policies. Farmers have a psychological preference for obtaining a 'fair' price, preferably from the market, for what they do rather than who they are. There is a welfare stigma associated with income transfers unrelated to production. There is little evidence that fully decoupled policies will ever be sufficiently acceptable to the traditional farm lobbies (or their political supporters or representatives) to replace existing support instruments completely. Second, decoupling focuses on reducing or eliminating production or supply-side distortions; it would not necessarily address demand-side or consumption distortions created by agricultural policies, unless decoupled instruments completely replace existing policies. Third, decoupling is a very general principle. Its implementation would require modification of existing policies on a case-by-case basis. There are no clear-cut criteria under which decoupled policies may be implemented by governments or monitored through an international body such as the GATT. [7]

PEG PRINCIPLES

It is against this background that the present proposal is made. The problem is to find a traditional commodity-based policy alternative which meets four basic criteria: a) minimises trade distortions, that is, generates production and

6   The argument that distortion and support are not the same thing is explored in de Gorter and Harvey (1990).

7   The proposal put forward by the US to the GATT negotiations for the replacement of all forms of agricultural support by their tariff equivalents and subsequent progressive reduction of these tariff equivalents - the 'tariffication' proposal - suffers from many of the same drawbacks. Direct income support and quantitative restrictions on production are discussed further in OECD (1990).

consumption levels which are as close as possible to those which would occur under free trade; b) achieves national farm support objectives; c) is politically acceptable to national governments; and d) is administratively feasible.

The principles which underlie the development of the Production Entitlement Guarantee (PEG) option are:

• that support of domestic agricultural industries is warranted by a variety of different objectives which cannot be subject to international negotiation. Among such objectives are the preservation of a rural structure which employs more people in agriculture than would be the case without support, and the provision of a safety net of support for a particular quantity of domestic production. This means that the focus of international agreement must be on trade distortion, not on levels of agricultural support;

• that support (protection) can be achieved without distorting world markets, where distortion is defined as the difference between domestic production or consumption levels under the policy from those levels which would occur without the policy. The essence of a non-distorting policy is that the 'incentive price' (the price on which production and consumption decisions are based) should be the free-trade world price;

• that development of non-distorting policies from those which are employed at present should be consistent with internal policy pressures and represent a minimal change from existing policy instruments. In other words, non-distorting policies should combine the alternative domestic responses of support price reduction (preferably to free trade world levels) and support control (through limits on production quantities eligible for support). Without this link, adjustment towards non-distorting policies will be politically impossible, since radical reform of domestic policies, even under the umbrella of international agreement, cannot be ratified through the political process.

A Production Entitlement Guarantee (PEG) is a pre-set fixed limit on the quantity of production eligible to receive support payments. Providing that this limited quantity is less than the quantity which would be produced without support, then the producers' incentive price is the market price, which partially satisfies the second principle. The second principle, however, implies that there should be no intervention which is not PEGed. The PEG requires the elimination of all existing border and internal agricultural support measures, except for payments on the specified PEG quantity. This means that consumers and users pay the free market price and farmers receive the free market price for any production in excess of the PEG quantity. It also eliminates the considerable expense and waste associated with the plethora of existing support mechanisms. The PEG limit should apply at both the national and the farm levels. Actual production is not controlled either at the national or the farm level. Farmers are free to decide how much to produce above the supported quantity.

PEG MECHANICS AND ANALYSIS

It is important that the PEG is set below production levels which would occur without market intervention, otherwise PEG payments will continue to distort international markets. The basic steps required for the introduction of a PEG scheme are as follows:

• establish a national PEG quantity upon which support payments are based that is less than the output that would be produced under multilateral free trade;

• distribute this PEG quantity among producers as their licence to receive support payments;

• eliminate all other border and domestic support measures so that domestic market prices equal world prices.

Figure 17.1 depicts the desired outcome under a PEG scheme for a particular commodity in a single country. The supply curve (S) shows how the current domestic support price (SP) generates production at quantity B. The existing world price is at WP0. Limiting support payments to the PEG quantity would initially lower production to that quantity. However, if all countries limit the production receiving support and remove all other forms of market intervention, world prices would rise, for example to WP1. In the diagram, the PEG quantity is less than the free-trade production level (A), and the marginal production (A-PEG) would be produced under competitive conditions at the world market price WP1. Therefore, a PEG set at a level below free trade output transfers income to farmers but results in no trade distortions. [8]

In fact, so long as farmers are not required to produce specific PEG quantities in order to receive payments, no trade distortion would result from a PEG which is set at a level greater than point A in Figure 17.1. This would make PEGs a fully decoupled programme and, as a result, could be less acceptable politically than the production-related PEG. However, PEG provides a mechanism which allows governments to move towards fully decoupled payments if they wish, although they need not do so in order to reduce trade distortions.

The Transferability and Issuing of PEGs

The likelihood of distortions increases if PEGs are non-transferable between farms or farmers. The reasons for this are illustrated by Figure 17.2, which

---

8   It may be argued that any coupled support, regardless of whether it is limited or not, will have some effect on the position of the supply curve, through its effect on the security of returns for instance. However, these arguments are complex and ambiguous, and the possible effects are considerably smaller than for unlimited support, particularly since the basis of PEG support is 'frozen' to historic production levels.

shows the marginal and average cost curves associated with three different farms. With an open-ended price support SP, farms 1, 2 and 3's production are at Q1, Q2 and Q3, respectively. Farm 1 is a high-cost farm. Under free trade it would produce nothing since the world price WP1 would be below its average cost of production. With PEG it will produce the amount P1, receiving price SP on the PEG output and more than covering costs. Farm 2, which is a medium-cost farm, would produce exactly the same quantity under the PEG as it would under free trade at world and market price WP1. The production decisions of low-cost farm 3 are undistorted by the implementation of the PEG scheme, since P3 is below the free trade quantity, and the expected production of farm 3 would be Q3', with support received only on quantity P3.

Figure 17.1: Diagrammatic Analysis of the PEG

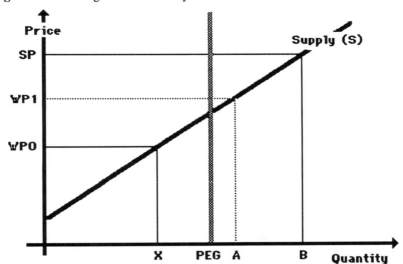

The implication of this analysis is that 'over-production' may occur with a PEG scheme even if the aggregate industry PEG is less than A in Figure 17.1. To overcome this problem, it would be necessary to a) permit the sale (rental) of PEGs between farms such that the most efficient farmers are able to purchase (rent) all of PEG1 and some of PEG2 (farm 1 would then go out of production and all output would be produced on the margin at the world price); and/or b) simply pay farmers a lump-sum PE-based payment regardless of their actual production levels (a fully decoupled payment). Their marginal production would then be determined by the world price. Without either a) or b) above, over-production and consequent world market distortion could result with the implementation of a PEG scheme. So long as governments agree to transferable PEGs or allow farmers not to produce the commodity to qualify for PEG payments, then the tendency for over-production would be minimised. An alternative is that the PEG quantity set at the national level should be lower if PEG licences are to be non-transferable than if they are to be freely tradable. [9]

Figure 17.2: Farm Level Analysis of the PEG

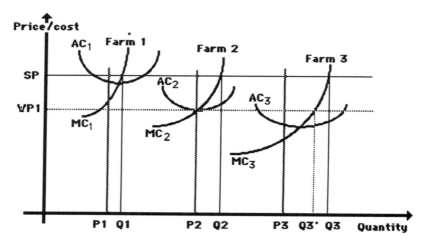

One of the major domestic advantages of PEG is that governments can maintain discretion over how and to whom the initial PEGs are issued. An argument can be made that PEGs should be transferable among farmers or farms on efficiency grounds but this may unduly limit domestic flexibility. To prevent an increase in enterprises (farms or farmers and commodity sectors) receiving support payments, countries could agree to bind the maximum level of support in each sector (using, for example, historical support levels) and bind the number of commodity sectors receiving support. This would limit potential trade distortions due to the effects of PEG payments on farmers' entry/exit, work/leisure and consumption/investment decisions. Individual PEGs could be issued to existing farmers on the basis of production quotas currently held in Canada and the European Community or land 'base' and 'programme' yields in the United States cereals sector. New entrants to farming would have either to purchase (lease) PEGs from other farmers or obtain them directly from the government; otherwise they would produce at world market prices and receive no PEG payments. PEGs could be issued by the government on the basis of the farm (in which case the value of PEGs will be reflected in the value of land) or for an individual farmer (in which case the value of PEGs will be reflected in the price of a certificate of eligibility).

9   Notice that fully decoupled PEG payments are **not** the same thing as requiring producers not to produce in order to receive payments, which does distort production decisions.

If governments wish to have a pool of PEGs available for new entrants, then a fixed percentage of all private PEG transactions (farm-to-farm sales, parents to son/daughter, and so on) could be automatically reclaimed by the government to pass on to these entrants. Indeed, such reclaimed PEGs could be used to reduce the aggregate level of PEGs to the left of A in Figure 17.1. Government purchases of PEGs could also be used to transfer PEGs to new entrants, to reduce the aggregate PEG level to ensure it is to the left of A, or to eliminate ('buy out') high-cost production which causes trade distortions. PEG buy-outs provide a mechanism to facilitate the exit of uncompetitive farmers and the rationalisation of the farming sector in the face of competitive pressures, while at the same time providing acceptable levels of compensation to those giving up the right to support. A rental market [10] would be an ideal method of transfer from an international perspective since this would permit the monitoring of whether a particular PEG results in a trade distortion. If the annual rental value of the PEG per unit is greater than the government's PEG payment, then production will be to the left of point A in Figure 17.1 and no trade distortion will exist.

Setting the PEG Level

The amount of trade distortion created by PEG depends upon where the PEG production level is set. For example, if historically a commodity was supported at SP in Figure 17.1 (with no quantitative restrictions), then output would have been at B. If world prices rise but remain below SP with the introduction of PEGs, then fixing PEGs at historical production levels would distort production and trade. If world prices rise above SP, as could be the case, then a PEG based on historical production will not be trade-distorting.

If PEG is to be a part of the reform of agricultural policies under the GATT, the quantity of production eligible for support must be determined as part of the negotiations. As an illustration, however, Figure 17.3 shows the estimated adjustment of current world prices towards free trade levels for selected commodities, with PEGs established at either 100 per cent of 1986 production (PEG100) or at 80 per cent (PEG80) of production. In both cases, the actual level of producer support per unit is kept at that actually estimated for 1986 through the Producer Surplus Equivalent (PSE), [11] with all other market intervention eliminated so that consumers pay market prices.

---

10 If the PEG payments provide some security for existing farmers, as they will, then the price of the PEG licences will include a premium for this security. As such, the indicator of 'excess support' suggested here will under-estimate the possible protection afforded by the PEG, though this is not expected to be a serious problem (see footnote 8).

11 As defined and calculated in OECD (1987). These estimates are based on the Roningen *et al.* (1987) analysis.

Figure 17.3: Percentage Adjustment of World Commodity Prices Towards Free Trade Levels Under PEGs at 100 per cent and 80 per cent of 1986 Quantities

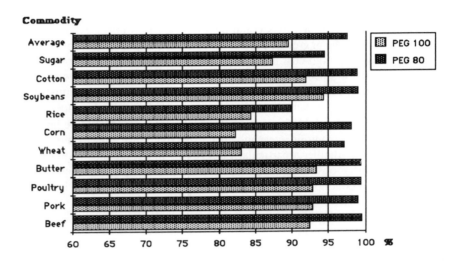

The results show that all sectors would experience at least 80 per cent of the free trade change in world prices under PEG100. On average, almost 90 per cent of the free-trade world price change would have occurred (top line in Figure 17.3). In the case of a PEG80 scheme, over 90 per cent of the full free-trade world price adjustment would have taken place in all cases, over 95 per cent of the full adjustment would occur for all but two commodities (rice and sugar) with an overall average of 98 per cent. Hence, a PEG of 80 per cent of 1986 production levels provides a rough indication of the appropriate goal for a negotiated PEG quantity if historical levels of support are maintained while still achieving the vast majority of the benefits of full liberalisation of world trade. [12] The results suggest that the PE limit on support needs to be lower than 80 per cent of 1986 production levels for sugar and rice, and more detailed analysis would reveal the different levels required in different countries for the PEG supports to be minimally trade-distorting.

12 These results come from recent empirical analysis by the International Agricultural Trade Research Consortium (IATRC) *Assessing the Benefits of Trade Liberalization,* (1988). This suggests that soybeans and soybean meal are the only exceptions to the general prediction that world prices would rise following trade liberalisation by industrial countries). Trade liberalisation would lead to an increase in world prices above current support prices for several commodities in several countries. Although the base year (1986) may seem rather out of date, the GATT negotiations are concerned with reductions in support from 1986 levels.

PEG IMPLICATIONS

PEG and Farmers

The PEG allows a threshold level of support to be paid to farmers on the basis of their production, recognising that there is a social benefit (implicit in current support policies) to the domestic production of farm products over and above the market price. The PEG payment can be viewed as the difference between the private (market) and social valuation of domestic production. The limit on this support recognises that this divergence abruptly diminishes as production quantities are increased. Furthermore, a part of this additional social value derives from the desire to support smaller 'family' farms for their contribution to the social and environmental fabric of the countryside, rather than the need to support large-scale 'industrial' agriculture. The distribution of PEG limits can be used to direct public support towards people rather than products and benefit smaller producers proportionately more than larger farms. In addition, conversion of present methods of support to PEG schemes would mean that farmers would get 100 per cent of the support directed towards them rather than only a fraction as under current policies. In particular, the waste associated with offsetting other commodity and countries' policies and with the implementation of market intervention mechanisms (including processing and storage) would be eliminated under the PEG option. Preliminary estimates suggest that this waste (the proportion of consumer and taxpayer expenditure on current farm support policies which does not get through to farmers) is about 60 per cent on average. While public dissatisfaction with the growth in this spending is almost all directed at agriculture, only some 40 per cent is actually benefitting farmers.

Figure 17.4 illustrates the extent to which world prices are estimated to increase for the major commodities. Although not strictly comparable, since the studies refer to different groups of countries and different assumptions about the behaviour of world markets, as well as different commodity coverages and base periods, the indications are that world prices would be appreciably improved as a result of multilateral liberalisation. On the basis of these estimates (specifically those of Roningen *et al.*1987) some 25 per cent of current spending on agricultural support around the world merely offsets the world price-depressing effects of other countries' policies and does not benefit domestic producers.

Implementation of the PEG option allows farmers to compete on a free and fair basis. Supply control is the only solution other than across-the-board cuts in support prices which can resolve the current difficulties for developed country agricultural policies. This option has already proved inevitable, if not attractive, to policy-makers for some commodities (notably dairy) and is likely to be introduced for other commodities, particularly cereals, in the future. Development of these policies, however, promises farmers increasing bureaucratic control over what they can produce and, in the light of increasing environmental concerns, how and where they can produce it. The future of agriculture under supply control is towards a public utility, with farmers as civil servants. It is difficult to believe that such a future is attractive to most farmers.

Figure 17.4: World Price Changes Following Industrialised Country Liberalisation (% Changes from Base Situations)

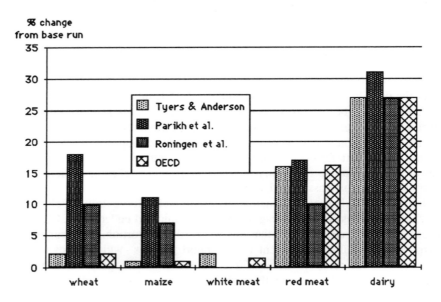

Source: Tyres and Anderson (1986); Parikh *et al.* (1986); Roningen *et al.* (1987); OECD (1987), adjusted to full liberalisation.

Key questions for farmers are: How big would the PEG limits per farm be? How much would the limited support payment per tonne be? What might the free-trade world prices be? As an example, consider the cereals regime in the European Community and the implications for the United Kingdom. Table 17.1 shows the dimensions of the possibilities assuming that the current level of taxpayer expenditure on cereals in the European Community is available to the United Kingdom on the basis of the UK's share of total cereal production.

In other words, if the European Community spends the same as it does now from the European budget on cereal support, and shares it out among countries on the basis of production, it could afford to pay £29/tonne over and above the market price on 400 tonnes per farm for every cereal farm in the UK. Estimates are that world cereal prices, which would be the market price for cereals in the UK, would be about £80 to £85/tonne under free trade. While the largest cereal producers in the United Kingdom would be disadvantaged under this proposal compared with the current situation, it is far from clear that they would lose compared with alternative developments of the European Community cereal regime, with more stringent application of the 'stabiliser' mechanism and the possibility of compulsory set-asides when support price reductions become politically unpalatable in the Council of Ministers. Given a reasonable phase-in period, it is likely that many competent cereal farmers in the United Kingdom (as well as elsewhere in the Community) could make money farming cereals under this set of prices.

Table 17.1: Country-Specific Cereal PEG: UK, 1986

Eligible production per farm (tonnes)	200	400	600	All
Coverage :				
- farms (%)	62	79	88	100
- production (%)	44	64	76	100
PEG payment*(£/tonne)	42	29	24	18

Source: Harvey and Hubbard (1988).
* Production-based budget share of £447.5 m (that is, 18% of £2.5 billion).

Phased introduction would be possible through gradual reduction of existing support measures and substitution of the PEG option as market prices in the EC fall towards (rising) world prices. Further phasing could involve the initial establishment of the PEG limit per farm at 100 per cent of base production levels and the progressive purchase of tradable PEG licences by the Intervention Authorities until the national PEG limit is achieved. In this way, farmers would be fairly compensated for voluntarily giving up the 'right to support'.

The programme would require farmers to establish their eligibility for payments on the basis of past production. It could also include, as part of the eligibility criteria, dependence on farming and on cereal production. In this case, either the amounts or limits could be increased. So long as the PEG licences were saleable, then efficient farms could afford to buy out the inefficient. Production patterns would not be frozen, and cereal production would tend to be concentrated in the most suitable areas. With upper limits on PEG ownership, concentration of support among a few could also be limited. The overall limit on the national proportion of production which can be supported under the PEG option would ensure that national supplies were not distorted by the PEG payments. [13]

Conversion of other commodity regimes under the Common Agricultural Policy in the EC is not difficult. The oilseeds sector already relies on crushing subsidies (aside from limited intervention in the seed market), which can be converted to limited production subsidies at the farm level. The EC sugar regime already involves the concept of limitations on the volume of production eligible for support. 'A' quota receives the full Community support price, 'B' quota is taxed with a co-responsibility levy, while 'C' quota sugar effectively receives the world sugar price, being taxed at the difference between domestic and world prices. Converting this system to the PEG involves elimination of border protection measures and the payment of a limited subsidy for 'A' quota to make up the difference between the domestic support price and the world price. Again, this shift would involve some increase in budgetary expenditure, but the increase would be reduced by the rise in world sugar prices following multilateral deregulation of the sugar market. The reform of the beef market within the Community would involve progressive replacement of the intervention mechanisms by a payment per head for breeding cows, on a limited basis per farm. Providing border protection and domestic intervention are also eliminated and the production levels eligible for support are kept within non-distorting bounds, this proposal is also consistent with PEG. Similar arrangements are possible for other EC products.

In the case of milk, the existing quota mechanism would be changed to a right to support rather than a right to produce. This right would be tradable between farmers within countries, and ideally between countries. Reductions in the amount of milk production eligible for support could be achieved by the intervention authorities buying in quota rights, rather than surplus milk products. Limits on the quantity eligible for support could be used to target aid to smaller milk producers without seriously distorting the pattern of production. Achieving the necessary changes on the demand side is straightforward in principle, but presents more problems for the political acceptability of the policy. In order to ensure that the incentive price for European Community consumers is the world price, intervention purchases of dairy products, export subsidies and import levies would need to be eliminated gradually. This would shift the burden of the support from the consumer to the taxpayer and would increase the budgetary cost of the programme. However, the change could be phased in as the production quotas were phased

13 The market price of cereals would fall by about 30 per cent. In this example, the proportion of production which would be covered by the PEG scheme would be 64 per cent of current production. For this PEG limit to be trade-distorting, it would require the elasticity of cereal supply in the UK to be greater than 1.2. This is substantially in excess of the majority of estimates of cereal supply elasticities, especially given that all other commodities are treated similarly, so that market prices for other commodities are also reduced.

out. Budgetary savings from the reduction in surplus disposal costs would be used to pay producer subsidies (on the PEG quantity).

Conversion of US policies to the PEG regime is beyond the scope of this chapter, except to point out that again there is no serious difficulty. For instance, the present cereal support policy in the United States relies on deficiency payments based on 'programme' yields associated with set-aside of base acreages. Both acreages and yields for this programme are already frozen under the 1985 Farm Security Act, which effectively PEGs the programme, although the PEG limit might not be acceptable. Further discussion of US policy liberalisation can be found in Gardner (1988).

Farmers will be concerned, rightly, about the stability of the free market and society has an interest in ensuring stability for an industry which relies on long production periods and significant (if currently excessive) capital investment. The potential instability of world markets is also a legitimate concern of consumers. While the subject deserves fuller treatment than can be provided here, several points are worth emphasising. First, as far as producers are concerned, the PEG payments provide a significant element of stability to most farm operations, if not to most production.

Second, a more liberal trading regime in the world market will improve the stability of the market substantially compared with the present situation. If the world market simply exists as a dumping ground for surpluses, as is the case at present, rather than as a competitive market for commercially viable supplies, then it is no surprise that world markets are unstable. An indication of the improvement in world market stability following liberalisation of world agricultural trade is given in Figure 17.5. Freely competitive markets provide incentives to producers, processors and distributors to improve stability through a variety of mechanisms, such as futures markets, crop insurance, storage and scheduling of purchases and deliveries. This is simply because buying products at low prices (that is, in times of surplus) and selling at high prices (that is, in times of shortage) is the classic recipe for commercial success. In so doing, supplies are rescheduled between periods of plenty to periods of shortage and the variation in prices is also reduced. Present farm policies, however, involve a substantial interference in both market prices and in stocks of commodities, at the whim of policy-makers. Neither encourage the commercial sector to try and 'play' the market to reduce instability.

Third, the stability of international markets is an international concern. It is not possible for a single country or country block to stabilise the world market on its own (except at unacceptable cost). In fact, as indicated in Figure 17.5, existing policies pursued to improve domestic stability actually destabilise world markets. The European Community's import levy/export refund system encourages exports and discourages imports as world prices fall below the European Community internal price, thus exacerbating the downswings in world prices. In those rare circumstances where European

Community internal prices are below world prices (as in 1974 for cereals) the reversal of the policy (export taxes) reduce exports exactly when the world is short of exportable supplies and prices are high. [14]

World market stability thus requires world (international) action and should be the concern of the GATT negotiations. However, a form of international agreement on stabilisation might include agreement about world price trends and variability, and acceptance that every country has both the right and the responsibility to stabilise domestic market prices around these agreed trends. Thus, it might be possible for the European Community to retain border intervention (export subsidies and import levies) so long as these instruments were used *only* to stabilise domestic EC prices around an agreed moving average of world prices. This means that export taxes and import subsidies would be used when world prices were below trend (encouraging imports and discouraging exports), and vice versa when world prices were above trend. Such action would tend to depress domestic prices when world markets were depressed and increase domestic prices when world prices were above trend. Stabilisation of the domestic market would then require counter-cyclical stock-holding policy by the domestic intervention authorities, building up public stocks when the market was depressed and running them down in times of surplus. A requirement that public stocks should be self-financing would prevent such activity being used as a back-door method of support.

Figure 17.5: Improvement in World Market Price Stability from Free Trade

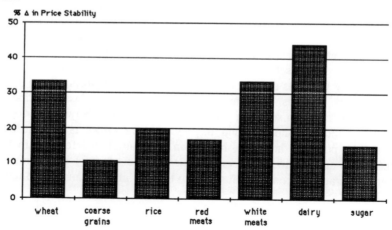

Source: Tyres and Anderson (1986).

14 This argument is straightforward for a single commodity, but, as noted by McCalla and Josling (1985, p 68), it is not clear that it holds in a multi-commodity world. However, the interactions between markets are incorporated in the models underlying Figure 17.5, which indicates that in practice domestic policies increase instability in the multi-commodity world market.

PEG and the Consumer and User

The implementation of a PEG involves the elimination of all other trade-distorting policy interventions which are used to support farm prices or incomes. Although the final consumer is clearly important (though often largely ignored in present developed country farm policies: see Chapter 7), the users of many farm products, especially grains, are farmers themselves. Reductions in the market prices of feedstuffs, following liberalisation of agriculture and the introduction of PEGs, will help farmers as well as consumers, and will remove the waste of having to support meat and livestock production simply because grains are supported.

In the case of the European Community, it has been estimated [15] that the consumer burden of the CAP compared with free trade is approximately equivalent to a 15 per cent Value Added Tax on food. Elimination of this tax can only benefit consumers. It might be objected that food security would be threatened by potential reductions in self-sufficiency ratios and greater dependence on imported foodstuffs. However, to argue this requires the specification of those conditions under which supplies from the world market would be interrupted. [16] There are essentially two sets of circumstances: a) trade wars and political dislocations; and b) armed conflict and physical prevention of trade. The present proposal is made specifically in the context of the GATT negotiations, which is aimed at minimising both the incentives and the ability of foreign suppliers to deny supplies to importing countries and at eliminating trade wars. It should be remembered that provision of access to domestic markets for foreign suppliers (a principal objective of the world's traditional agricultural exporters in the GATT negotiations) carries with it a responsibility to ensure adequate supplies to those importers. This obligation can and should be enshrined in the GATT provisions dealing with agriculture.

Security of food supplies in the case of armed conflict (the historical reason in Europe for concern over this objective) need serious reassessment in the light of modern agriculture and modern warfare. First, modern agriculture in Europe is heavily dependent on oil-based inputs. Whether imported or domestically produced, these supplies are extremely vulnerable to dislocation and interruption in times of war, which would make a mockery of existing levels of food self-supply. Second, the supply of home-grown foodstuffs in the event of either chemical or nuclear attack is obviously totally insecure!

As a final comment, however, freer trade will not prevent chronic and acute food shortages occurring from time to time around the world, particularly in parts of the developing world. It is hardly accidental that the industrialised world also includes those areas of the planet which are more

---

15 Harvey (1982, 1984) and Harvey and Thomson (1985) also report estimates and implications of the costs and benefits of the CAP.
16 The issue of food self-sufficiency and security is dealt with by Ritson (1980) and will not be repeated here.

suitable for reliable food production without large-scale irrigation or drainage requirements. The twin problems of providing improved stability of world prices and increased food security for the whole planet are essentially international problems. They can only be satisfactorily resolved at the international level and they deserve international attention, potentially under the GATT.

PEG and the Taxpayer

A major political disadvantage is that all transfers under PEG are paid by taxpayers and hence are more visible. Furthermore, much of the domestic pressure on existing farm policies on both sides of the Atlantic comes from their excessive budgetary cost. A proposal which promises to increase this spending rather than reduce it cannot hope to be taken seriously. Maintaining a desired level of producer income by switching the entire cost of support to taxpayers could increase government expenditures substantially for some commodities, especially in Japan and the European Community. However, there are several features of a PEG option which substantially reduce these problems.

• Co-ordinated international action to deregulate agriculture will increase world market prices. Current estimates are that as much as 25 per cent of current spending on agricultural support is wasted overcoming the price-depressing effects of other countries' policies. So the gap between the world market price and a 'fair' PEG price will not be as large as the current differentials between domestic support prices and world prices. The resulting taxpayer savings can be used to replace traditional transfers from consumers in order to maintain producer income (indeed, total taxpayer costs could decline in some cases).

• The efficiency of income transfer improves, since much of the current transfers, especially under EC and US programmes, are wasted (given to importers) or are 'deadweight losses' (spent on processing, storage and so on) and never reach domestic farmers. As mentioned earlier, it is estimated that another 35 per cent of support expenditure leaks away through these drains, making 60 per cent waste in all. In other words, only 40p of each £ paid by consumers and taxpayers for farm commodity support actually ends up in farmers' pockets. With a PEG scheme, all the taxpayer support is transferred directly to farmers as income, with little loss due to 'over-production' and transfers to the rest of the world. Furthermore, a PEG scheme will benefit livestock, poultry and dairy farmers in many countries by reducing cereal prices, and thus reduce the extent to which these producers require support.

• Taxpayers are consumers: all that a PEG scheme does is alter the method by which income transfers to farmers are made, so that the net effect on households, even with an increase in budgetary cost, cannot be negative (providing that support levels are not increased). Using high consumer prices to support agricultural incomes is also generally more regressive than using tax revenues, since it places the burden of paying for support on the poor, who spend a greater proportion of their income on food than the rich.

• If traditional levels of producer income cannot be maintained without unacceptable increases in taxpayer costs, then governments can limit the per farm transfer to keep within budget constraints. Such targeting could be used to assist small or family farms and disadvantaged areas, rather than providing support for larger or richer farmers. In any event, the PEG option provides more scope for 'cash-limiting' farm support costs than existing programmes, including all those employed by the EC.

As a result, although the PEG option may at first sight appear to be ruled out by the potential increase in budget spending, closer analysis reveals this fear to be misplaced. As already illustrated by the EC cereals example above, it is possible to design a PEG scheme which maintains the present level of support for the vast majority of farmers at existing levels of taxpayer spending. It may be argued that these levels are excessive. It is not the purpose of this proposal to discuss that question, which is a matter for individual countries or trading blocs to resolve. However, if the political process decides that support should be reduced, the tradable PEG option provides a mechanism for doing so, simultaneously providing acceptable compensation for those who voluntarily give up the right to support, simply through the authorities buying up PEG licences.

PEG and the Administrator

Any system of agricultural support which decouples support payments from production or sales faces problems of administration. Those associated with the PEG are no more difficult than any other decoupled scheme, and potentially much easier than some other options, such as negative income taxes or welfare payments. PEG is at least as easy to operate as supply management schemes, particularly since farmers are paid only if they follow the rules under PEG (with the penalty of exclusion from the programme for rule-breakers) rather than being forced to follow the rules with no direct benefit as is the case with supply management procedures.

It is difficult to believe that an administration which is capable of implementing intervention buying, quota arrangements, payments and premiums based on numbers of cattle or olive trees and so on (as is currently the case under the CAP) is incapable of administering a programme which pays farmers a specified amount based on the production of an authenticated PEG certificate. The initial allocation of PEG certificates is a potential problem, but again the fact that possession is a support entitlement rather than a production licence should make for more co-operation from farmers, and this is a 'once-and-for-all' cost, not an ongoing difficulty.

PEG and the Political Process

No doubt farmer organisations and farm ministers will find the idea of PEGs difficult to accept in the first instance. However, the commonly accepted objectives of farm policy are all met at least as well by the PEG system as by current farm policies. The political process can be seen as matching political objectives (reflecting social requirements for the agricultural industry and the countryside) with policy instruments and settings. In most cases, political objectives differ by region and local circumstance. These differences are very imperfectly catered for under a policy of market price intervention, which provides uniform price support for all. The PEG option provides flexibility for tailoring support levels and distribution to provide support for those deemed to be in most need or most deserving. It is consistent with, and represents a practical transition to, a system of direct income support for agriculture.

Direct income supports have been advocated by the European Commission for some time, and have finally been agreed by the Council of Ministers in 'a potentially far reaching change to the Community's traditional methods of subsidising agriculture'. [17] Reports of this agreement, by majority vote with the UK against, suggest that it will trigger fears of incipient fragmentation - or renationalisation - of the CAP. The UK position is reported as disagreement with the whole scheme [18] 'which, to us, is really just a social security scheme.' However, this position seems to ignore both the practical impossibility of gaining political support for the UK's preferred solution (dramatic reduction of support prices with no proposals for alternative support mechanisms) and the considerable arguments in favour of retaining some support for the rural areas and for a base level of agriculture within these areas. Even allowing that the current level of support is too high, no matter how paid, there is a requirement for some mechanism for reducing support while providing compensation and assistance to those most in need. The PEG option, combined with progressive reduction in PEG quantities and associated licences, provides such a mechanism. It would be an act of considerable political leadership within the EC for the UK to adopt and promote such a policy option.

Those who stand to lose from the change are the larger farmers, who currently receive the lion's share of agricultural support. Not only are these farmers generally held to be the more efficient, and thus should be able to cope

---

17 *Independent* 25.1.89. The arguments have been repeated as the same proposal is being discussed in January 1990, in a last attempt to salvage the Uruguay round and the CAP budget crisis.

18 Ibid. The scheme is to be voluntary for countries but 'other EC countries, notably West Germany, are certain to press ahead'. Three-quarters of the support is to come from member states with Brussels contributing up to £700/farm and a ceiling of £1,050/farm on national additions. Sharing of funding is to be skewed towards EC in the case of poorer member states.

with a fair and level market-place, but also they hardly constitute a legitimate target for income support payments. Nevertheless, it should be recognised that it is these vested interests who have most to gain from a perpetuation of the present system, even though the logical conclusion of present policies in the European Community is increasing central control over what is produced and (in the light of the growing environmental pressure) who shall produce it and how. Consumers, users, taxpayers and the vast majority of farmers, especially the small and disadvantaged, all stand to gain from a change to PEG policies. In that sense, political unacceptability is an indictment of the so-called democratic system which makes the political decisions.

Proposals that involve a switch from production-based support to a system of direct payments, of which PEG is one, raise issues of national interest within the Community, which often seem to dwarf Community interest. For example, it appears that one reason for British reluctance to embrace a major switch in the direction of the CAP support towards 'small farms' is that Britain, having by continental standards very few small farms, would lose out. The logic of this argument is, however, spurious. If it is accepted that small farms need support if they are to survive, and that their survival is critical for the conservation of rural natural, social and economic environments, it certainly does not follow that the appropriate definition of 'small' applies uniformly throughout the Community. In fact, the size and type of farm which would be vulnerable under a free market system yet crucial to the conservation of the countryside must vary substantially by region, let alone by country. It stands to reason that the definition of small cannot be 'common', and that it must be determined on a regional, and hence national, basis.

How should such support be funded? Common financing requires that the European budget contribute to nationally determined support mechanisms within a commonly agreed European framework. Why not simply fix a ceiling on European budget support at the agreed financial ceiling on FEOGA market support spending and divide this total between member states on the basis of current production shares? Member states could, under certain specified conditions, be free to supplement this support from national funds, providing that all support payments were completely 'decoupled' from production levels and decisions. Efficient farmers would then be free to compete on a level playing field for market shares within and outside the Community. How could Britain, or indeed any other member state, lose out from such an agreement?

PEG and the Environment

It would be foolish to argue that the PEG option provides a solution to the multitude of rural environmental problems currently causing concern in most developed countries. However, part of these problems arise because current unlimited levels of support have encouraged an excessive intensification and capitalisation of agriculture with consequent problems for the natural environment. Furthermore, the fact that support based on levels of output provides greater benefits to larger farmers encourages increases in the size of farms, and associated increases in the size of fields and loss of field boundaries, which are often a major ecological asset. Fewer and larger farms typically employ fewer people and more capital per hectare, and offer more limited opportunities for part-time or complementary activities. None of these effects are environmentally friendly, while all are encouraged by present support systems.

Conversion to PEG would provide opportunities to reverse these tendencies and to combine more specific environmental policies with farm support programmes. Furthermore, so long as the PEG licences are independent of land or other agricultural assets, land prices will fall as a result of this change. Alternative land uses, for environmental purposes, can then compete more effectively with agriculture and the process of environmental improvement is therefore made easier and less expensive. For example, the European Community's set-aside scheme, which pays farmers to leave land idle (or convert land to other uses) rather than grow cereals, currently has to compete with the high support price for cereals. This nonsense is entirely removed by the use of the PEG option as outlined above.

PEG and the GATT

There are a number of possibilities for including the PEG option directly within the GATT process. At one extreme, the PEG option may simply serve as a practical example of an essentially decoupled, non-trade distorting policy instrument which would be encouraged but not required. At the other extreme, PEGs could be treated as the agricultural equivalent of the tariff. GATT agreements would involve commitments to convert all existing agricultural policies to their PEG equivalents; binding of negotiated PEGs and PEG payments under the GATT; progressive reductions of PEGs and monitoring of effectiveness through the trading price of PEGs compared with PEG payments (as the difference between support prices and world (domestic) market prices). It might even be suggested that disputes between countries about the appropriate levels of PEGs could be resolved (with appropriate international compensation) through one nation reducing PEG levels of competitors by purchasing the 'offender's' PEGs, thus effectively reducing the supported quantities! [19] It is beyond the scope of this chapter to examine these alternatives in detail. However, the proposal does not restrict the GATT process to include the PEG option in any particular form. Indeed, it is a primary

---

19 There would, of course, be no question under this option of the purchasing country being entitled to the selling country's support payments on the purchased PEGs. They would simply become inoperative, as with a domestic authority's purchase.

purpose of the proposal to help the GATT negotiations and to further progress towards a more liberal agricultural trading regime.

CONCLUSIONS

Governments intervene in agricultural markets to achieve farm income support objectives. The role of the GATT negotiations is to minimise international trade distortions resulting from these measures, not necessarily to reduce 'protection' to the farm sector. In order to achieve less distorting farm income support, this chapter advocates a measure called 'Production Entitlement Guarantees' (PEGs). This limits the quantity of production eligible for support at the farm level and replaces all other forms of direct or indirect income support to farmers. The advantages of PEG are that it:

• provides a means for governments to reduce trade distortions while maintaining farm incomes;

• allows countries to realise mutual gains from freer trade through increased world prices and reduced consumer prices;

• involves a known and therefore limited level of budgetary expenditures;

• is consistent with traditional commodity programmes which provide support to farmers based on production, and therefore is a realistic alternative to existing programmes;

• is a more cost-efficient mechanism for transferring income to farmers than current agricultural programmes;

• provides substantial national flexibility in the targeting of support in terms of commodities, farms, farmers or regions;

• is consistent with recent trends in limiting support payments (through quotas) in many countries;

• can provide a mechanism for transitory compensation and adjustment if desired;

• is easily implemented at the national level, as a minor change to existing programmes;

• lends itself to negotiation and monitoring through the GATT;

• enhances the world's international trading relations and reduces the threat of agricultural trade conflicts.

It is argued here that a principal deterrent to the negotiation of a more liberal trading system for agriculture is the lack of any clear alternative policies which are less trade-distorting and which are politically acceptable at home. The PEG option is one such alternative. Perhaps it can be used to assist the vital task of removing the insanity and obscenity of the present agricultural trading system.

As this book goes to press, the process of liberalisation of Eastern Europe is gathering momentum, with the historic reunification of Germany leading the way. It is a stated intention of many newly democratic Eastern European countries to join the European Community as soon as possible. It is clear that the present protectionist CAP cannot absorb actual and potential agricultural exporters without radical dislocation of the policy. Yet it is also clear that both within the present EC and in potential new entrants there continues to be strong political and social pressure for continued support of the agricultural sector. It would be a disaster if the reunification and liberalisation of Europe were to founder on the rock of an outdated, inefficient and ineffective agricultural policy. The proposal advanced here is capable of resolving these difficulties. For that reason, if no other, it deserves serious consideration.

REFERENCES

de Gorter, H. and Harvey, D.R. (1990) *Agricultural Policies and the GATT: Reconciling Protection, Support and Distortion,* Paper to EAAE Congress, The Hague, September.

Gardner, B. (1988) *Domestic Policies to Make Trade Liberalisation Politically Possible: the US Case,* Background paper prepared for the IATRC Task Force on Designing Acceptable Agricultural Policies, chaired by D. Blandford, Cornell University, Ithaca, NY, USA.

GATT (1988) *News of the Uruguay Round: Montreal Meeting of the Trade Negotiations Committee,* NUR 023, December 14, Information and Media Relations Division, Geneva.

Harvey, D.R.(1982) National Interests and the CAP, *Food Policy,* 7(3), pp 174-190.

Harvey, D.R. (1984) The CAP: Continued Agricultural Protection or Curb Agricultural Production, Paper II, *Whither the Common Agricultural Policy of the European Community?,* University of Newcastle upon Tyne Agricultural Society.

Harvey, D.R. (1988) *Food Mountains and Famines: the Economics of Agricultural Policies,* DP 5/88, Department of Agricultural Economics and Food Marketing, University of Newcastle upon Tyne.

Harvey, D.R and Thomson, K.J. (1985) Costs, Benefits and the Future of the Common Agricultural Policy, *Journal of Common Market Studies,* 24(1), pp 1-19.

Hubbard, L.J. and Harvey, D.R. (1988) *Limited Support Payments: an Option for the EC Cereal Regime,* Discussion Paper DP 4/88, Department of Agricultural Economics and Food Marketing, University of Newcastle upon Tyne.

International Agricultural Trade Research Consortium (IATRC) (1988) *Assessing the Benefits of Trade Liberalization*, Summary Report, IATRC, Ithaca, New York.

McCalla, A.F. and Josling, T.E. (1985) *Agricultural Policies and World Markets*, Macmillan, New York.

OECD (1987) *National Policies and Agricultural Trade*, Organisation of Economic Co-operation and Development, Paris.

OECD (1990) *Reforming Agricultural Policies: Quantitative Restrictions on Production; Direct Income Support*, Organisation of Economic Co-operation and Development, Paris.

Parikh, K.S, Fischer, G., Frohberg, K. and Gulbrandson, O. (1986) *Towards Free Trade in Agriculture*, Food and Agriculture Programme, International Institute for Applied Systems Analysis, Geneva.

Ritson, C. (1980) *Food Security and Self-Sufficiency*, CAS Paper No. 8, Centre for Agricultural Strategy, University of Reading.

Roningen, V., Sullivan, J. and Wainion, J. (1987) The Impact of the Removal of Support to Agriculture in Developed Countries, Paper presented at American Agricultural Economics Association meeting, East Lansing, Michigan, USA.

Tyres, R. and Anderson, K. (1986) Distortions in World Food Markets: a Quantitative Assessment, Background Paper for the World Bank's *World Development Report*, Washington.

POSTSCRIPT

Having spent a considerable part of his early career as a civil servant in the Ministry of Agriculture, John Ashton was acutely aware of the problems and difficulties of injecting rationality into the policy process. While this did not discourage him from trying, he also placed great store on the understanding of both the effects of policies and the reasons for their survival, often in the face of considerable pressure. To quote him, [1] 'I sometimes wonder in fact if agricultural economists generally are as aware as they should be of the diverse interests that influence agricultural policy and that all of these interests are worthy of study by agricultural economists.' These essays continue that tradition and provide a systematic and wide-ranging treatment of the European Community's Common Agricultural Policy.

One of the major contributions which John engineered on his arrival at Newcastle was the establishment of the Agricultural Adjustment Unit, funded through a generous grant from the Kellogg Foundation. The central focus of this Unit was the combination of understanding the processes and opportunities of the adjustment of agriculture and of the extension of this information and knowledge to the agricultural community. Coinciding with this programme, a major attempt was made to model UK agriculture so as to provide reliable projections of the effects of the dramatic change in the system of support, consequent on the UK's accession to the European Community and the Common Agricultural Policy. This work was sponsored by the MAFF and others.

Later, the Department was commissioned by the MAFF to develop a model to determine the distribution of the costs and benefits of the CAP among the member states and between consumers, producers and taxpayers. The resulting CAP model (discussed briefly in Chapter 5) subsequently developed under an SSRC grant, has proved successful in providing answers which have been judged useful by a variety of different institutions and individuals. These two illustrations demonstrate John's commitment to using the tools of neoclassical economic analysis to provide useful and credible answers to policy questions. 'Academic' questions seldom interested him; the policy purpose provided the spur to academic development and advancement.

It is tempting to characterise current circumstances as marking a turning-point in the development of the CAP. The coincidence of the GATT negotiations on levels of agricultural protection around the world with the preparations for the Single European Market to be born in January 1993, the

---

1   Presidential address, EAAE, reprinted at the beginning of this collection of essays.

continued growth of environmental concerns and policies which attempt to deal with them, and more recently the remarkable disintegration of central planning autocracies East of the now rusty iron curtain, can all be interpreted as providing the basis for a thorough going reform of the policy. Nor is there any shortage of suggestions about how this might be done. But the history of the policy does not suggest that reform will be either sudden or radical. A process of evolution rather than revolution is a more likely characterisation.

John Ashton was fond of the analogy between policy proposals and London buses: 'Wait a few moments and another one will come along.' Students of the London bus fleet will notice some simple extensions of this analogy: first that the policy options which do come along are nearly all at least 30 years old, albeit with many reconditioned parts. Second, the frequency of arrival is in inverse proportion to need: they are never around when you want them, but are queuing up when you have time to spare. Third, except by the greatest good fortune, no one bus is able to take you exactly from your starting-point to your final destination: as pointed out in Chapter 16, we almost certainly need at least as many buses (policy instruments) as we have destinations (objectives) in order to satisfy our travellers. Certainly the basic policy options of price reductions versus supply control, and its juvenile form - producer co-responsibility - have been around for years, but neither have got us to our destination yet.

In the light of this apt analogy we would be the last to claim either originality or universality for the proposal in the last chapter - the PEG - which has clear antecedents in both the French quantum system and the British Standard Quantities associated with Deficiency Payments. Yet the proposal continues a tradition of constructive policy analysis firmly established by John Ashton at Newcastle.

The chapters of this book represent, we believe, a fitting memorial to the interest and enthusiasm which John generated amongst both colleagues and students. We hope that he would feel that our efforts do him justice. The Common Agricultural Policy has been of concern to all interested in agriculture since its inception, during which time there have been a number of 'crises' which have tempted some of us to believe that radical reform was imminent. John cautioned against this temptation, arguing strongly and cogently that policies develop their own special inertia and vested interests which usually prove sufficiently strong to resist seemingly irresistible pressure for change. Once again, the current pressures, especially through the GATT, and stemming from the liberalisation and democratisation of Eastern Europe, offer both strong reason and great opportunity for real, if not radical, change. It would please John more than anything if these pressures and opportunities finally bore fruit.

As this book goes to press, the University of Newcastle upon Tyne is releasing news of an extremely successful appeal in memory of its late Chancellor. This will be used to fund the endowment of the 'Duke of Northumberland Chair of Rural Economy', together with the establishment of a Centre for Rural Economy, in the Department of Agricultural Economics and Food Marketing. Just as the Agricultural Adjustment Unit reflected the policy pressures facing British agriculture during the 1960s, so the new Centre will focus on current concern over the relationship between farming, the countryside and the environment, continuing the tradition begun by John Ashton of major research initiatives at Newcastle University directed towards applying the theory of agricultural economics to contemporary rural problems.

Index

revenue-increasing measures 306–8
US views on 269–71
research and development
EC involvement 200–2
expenditure, UK 196–8
funding 191
responsibility for 185–6
'Rotterdam effect' 282
rural areas
decline 158
development advice 161
identification 141–2
standard problems 158

SCA 29, 102, 107
SCAR programme 200, 201
second-stage processing 208, 212–14
'self-financing' product levies 306
self-sufficiency 131, 133
set-aside scheme 85, 195, 305–6
US views on 270–1
single European market 165–6
agricultural trade implications 179–81
and food industry 216–17
Smithsonian Agreement 49
soybean, US exports to EC 265–7
Spaak Report 24–5
Spain
implications of entry to EC 64–6
intra-EC agricultural trade 176
Special Committee on Agriculture 29, 102, 107
Stabex 254
stabilisers programme 270, 304
strategic models 91, 92
Stresa Conference 27–8
structural funds, objectives 160
sugar
preferential trading agreements 253–4
price support 209, 211
surpluses 5–6
dealing with 78–9
destruction 120, 127, 128–9
see also intervention measures
switchover mechanisms 180

tactical models 91, 92
target price 4
technological change 186–8
and the CAP 194–6, 198–200
consequences of 191–4
modelling 188–91
valuations 199
threshold price 4
trade
barriers to 166, 180–1
creation 167–8, 288–9
diversion 167–8, 289
EC 172–3, 228–31
effect of customs union on 167–9
government restrictions 166–7
theory 166–7
world, and the CAP 223–8, 242–3
see also agricultural trade
Treaty of Rome 1–3, 25–7, 207–8
Tunisia, EC agreements with 278

Unit of Account (UA) 6, 47–8
United Kingdom
agricultural policy and CAP 72–86
agricultural research and development
expenditure 196–8
changes in agricultural labour 148
changes in agricultural land use 146–7
conflicts of interest 112
entry to EC 50
estimated cost of accession to CAP 95–6
forestry, impact of CAP on 151–3
implications of adopting CAP 51–2
intra-EC agricultural trade 176
United States
and CAP reform 269–71
perception of GATT talks 271–4
Super 301 countries 259
trade developments with EC 260–4
trade problems with EC 265–9
views on CAP 264–5

wheat, world market instability 246–52
world prices 124, 125

'zero' option 272